CLIMATE CHANGE

Insights You Need from Harvard Business Review

Business is changing. Will you adapt or be left behind?

Get up to speed and deepen your understanding of the topics that are shaping your company's future with the **Insights You Need from Harvard Business Review** series. Featuring HBR's smartest thinking on fast-moving issues—blockchain, cybersecurity, AI, and more—each book provides the foundational introduction and practical case studies your organization needs to compete today and collects the best research, interviews, and analysis to get it ready for tomorrow.

You can't afford to ignore how these issues will transform the landscape of business and society. The Insights You Need series will help you grasp these critical ideas—and prepare you and your company for the future.

Books in the series includes:

Agile

Artificial Intelligence

Blockchain

Climate Change

Coronavirus: Leadership and Recovery

Customer Data and Privacy

Cybersecurity

Monopolies and Tech Giants

Strategic Analytics

The Year in Tech, 2021

CLIMATE CHANGE

Harvard Business Review Press
Boston, Massachusetts

HBR Press Quantity Sales Discounts

Harvard Business Review Press titles are available at significant quantity discounts when purchased in bulk for client gifts, sales promotions, and premiums. Special editions, including books with corporate logos, customized covers, and letters from the company or CEO printed in the front matter, as well as excerpts of existing books, can also be created in large quantities for special needs.

For details and discount information for both print and ebook formats, contact booksales@harvardbusiness.org, tel. 800-988-0886, or www.hbr.org/bulksales.

Printed in the United States of America

10 9 8 7 6 5 4 3 2 1

Library of Congress Cataloging-in-Publication Data

Names: Harvard Business Review Press.
Title: Climate change : the insights you need from Harvard Business Review.
Other titles: Climate change (Harvard Business Review)
Description: Boston, Massachusetts : Harvard Business Review Press, [2020] | Series: Insights you need | Includes index.
Identifiers: LCCN 2020013972 (print) | LCCN 2020013973 (ebook)
Subjects: LCSH: Climatic changes—Economic aspects. | Social responsibility of business.
Classification: LCC QC903 .C552445 2020 (print) | LCC QC903 (ebook) | DDC 333.71/3—dc23
LC record available at https://lccn.loc.gov/2020013972
LC ebook record available at https://lccn.loc.gov/2020013973

ISBN: 978-1-63369-992-2
eISBN: 978-1-63369-993-9

The paper used in this publication meets the requirements of the American National Standard for Permanence of Paper for Publications and Documents in Libraries and Archives Z39.48-1992.

Contents

Introduction

CLIMATE ACTION: THE NEW BUSINESS IMPERATIVE

by Dante Disparte

Climate change demands immediate action—from organizations, governments, and society alike. But business leaders often find themselves asking, *how*? How should companies respond? What steps are effective given operating and financial constraints? What can realistically move the needle for an issue so vast? Will shareholders and other stakeholders forgive potentially inferior returns in favor of global longevity? This last question underscores the fundamental tension between those organizations that want as much as they can have as fast as they can have it and those who want to stretch the economically viable horizon for as long as possible. Climate change should push every leader, every industry, every country, and every person to aim for long-term prosperity.

Against these questions, responses to climate change have often erred on the side of incrementalism—or, worse yet, token steps bordering on "greenwashing" or marketing campaigns. There is good news, though. First, there are real measures leaders can put in place to position their firms (and collectively, the planet) on a path to equilibrium, resilience, and competitive advantage. Second, stakeholders can see through the thin veneer of greenwashing and expect enterprises to rise to the occasion.

In the last 20 years, there have been more disaster declarations in the United States than the preceding 50. The cluster of multibillion-dollar disasters, including Hurricanes Katrina, Sandy, Maria, Irma, and Harvey, were in concentrated areas, showing the link between extreme weather and rising temperatures. Hurricane Harvey, for instance, was the wettest storm in U.S. history, turning Houston, Texas, into a lake. Hurricane Maria triggered the second-longest blackout in history, plunging the U.S. territory of Puerto Rico into six months of darkness and claiming more than 3,000 lives. California's fire seasons, meanwhile, courtesy of a progressively hotter and drier environment, have worsened each year, with the 2018 Camp Fire being the deadliest and costliest on record, triggering what may be the first climate change bankruptcy of the electric utility, PG&E.

The same pattern holds true around the world, with unprecedented speed and scale. Typhoon Haiyan in the Philippines caused the worst recorded blackout in history, showing how critical infrastructure, as currently designed, can be a single point of failure. Australia's record 2019 bushfires caused devastation across the continent, killing over a billion animals and causing some species to go extinct. The last decade has been the hottest in history, which creates an environment where vector-borne diseases, like Zika, will continue their inexorable march north. The threat of communicable diseases and pandemics, like COVID-19, will also thrive in a world where climate change goes unchecked. While climate science and mitigation efforts often focus on large-scale impacts, climate risk is both attritional and acute when it comes to critical infrastructure that keeps the global economy going.

In the face of such threats and catastrophes, the private sector has the wherewithal, ingenuity, and balance sheet to make a material difference. Some businesses are already leading the charge. Clothing companies like Patagonia and H&M have been encouraging customers to use less by repairing clothing, embracing a business model of degrowth. Investors and banks, like HSBC, are looking to offer deals and funding to companies that emphasize sustainability. Other companies have made carbon reduction a priority—and partner with suppliers that do the same—in order to align themselves with the Paris Agreement and reach net-zero emissions by 2050.

Organizations must follow these examples for the sake of the earth and future generations. There is opportunity in responding to climate risk, and leaders must have the clarity to make the right choices to position their firms for longevity.

What This Book Will Do

This book addresses this opportunity in an accessible and practical way. It offers pragmatic guidance to help shape strategy, company responses, and approaches to climate action that all business leaders can apply.

The first section of the book will help you understand the key drivers of climate change, its causes, and its global impacts. It lays out the leadership priority of investing in the new economy while divesting from the carbon-hungry economy of the industrial age. You will learn how looser environmental regulations across the globe contribute to more pollution, even from those companies reducing emissions globally, and you'll also see the state of responsible investment and asset allocation and how ESG–focused investment is becoming mainstream.

Rising to the climate change occasion requires bold moves, institutional fortitude, and clear-sighted leadership. The second section outlines key considerations for reducing adverse climate impacts, responding to climate change in the present tense, and moving away from incremental steps to more holistic, whole-of-business strategies, from folding carbon-reduction into your everyday operations to building organizational resilience in the face of catastrophe.

As you read this book, consider whether you and your organization are ready to make the important commitment toward immediate climate action by asking some key questions:

1. How is my company's investment and operational strategy aligned to the reality of climate change?

2. Is my organization ready for a zero-carbon future and ready to respond to the emerging pressure of stakeholders, employees, and customers? If not, what steps do we need to take to become ready?

3. What new business models or innovative products can my company pursue to reduce emissions,

encourage dematerialization, and contribute to degrowth?

4. Should catastrophe strike, how will my organization respond to protect people, communities, supply chains, and other assets? Are we prepared for such instances?

5. What policies should I, as a leader, be backing, and how can I lobby these policies? Who should I be speaking to, and how can I make my commitment to climate action known to the public?

Much can be done to fight climate change, and it is up to organizations to join the charge. Leaders must make resilience, equilibrium, and true stakeholder capitalism a way of life. Reframing climate change as a business imperative and unique market opportunity is a necessary part of this change. The time is now. Actively work to prevent climate change and pave the way for a better future.

Section 1

UNDERSTANDING CLIMATE CHANGE

LEADING A NEW ERA OF CLIMATE ACTION

by Andrew Winston

Climate change is a global emergency. It's threatening crops, water supplies, infrastructure, and livelihoods. It's damaging the broader economy and company bottom lines *today,* not in some distant future. In recent years, AT&T has spent $874 million on repairs after natural disasters that the company ties to climate change. The reinsurance leader Swiss Re has seen large increases in payouts for damage caused by extreme weather events—$2.5 billion more in 2017 than it had predicted—a trend that CEO Christian Mumenthaler attributes to rising global temperatures.[1] If we don't move quickly toward action on climate, says Mark Carney, the Bank of England governor, we'll see company bankruptcies and raise the odds of systemic economic collapse.[2]

Corporate leaders are at last absorbing this; nearly every large company has significant plans to cut carbon emissions and is acting. But given the scale of the crisis and the pace at which it's developing, these efforts are woefully inadequate. Critical UN reports in 2018 and 2019 make two things clear: (1) To avoid *some* of the worst outcomes of climate change, the world must cut carbon emissions by 45% by 2030 and eliminate them entirely by midcentury.[3] (2) Current government plans and commitments are not remotely close to putting us on that path. Emissions are still rising. (See figure 1-1.)

Countries, cities, and businesses need to move simultaneously along two paths: reducing emissions dramatically (mitigation) and investing in resilience while planning for vast change (adaptation). My focus here is on mitigation, because adaptation alone—building ever-higher walls to keep out the sea and simply turning up the air-conditioning as the outdoors becomes uninhabitable—won't save us. If we allow climate change to destroy the plant and animal ecosystems we rely on, there will be no replacements. The good news is that business has enormous potential to profitably cut emissions faster and even more.

FIGURE 1-1

Alarming forecast: current climate policies are grossly inadequate

To hold global warming to 1.5° Celsius above preindustrial levels and prevent the worst impacts of climate change, the world must cut carbon emissions to zero by midcentury. Yet emissions are still rising, and under existing policies reductions won't begin to approach what's needed. If we stay on the current path, temperatures will probably increase by about 3° C, with catastrophic effects.

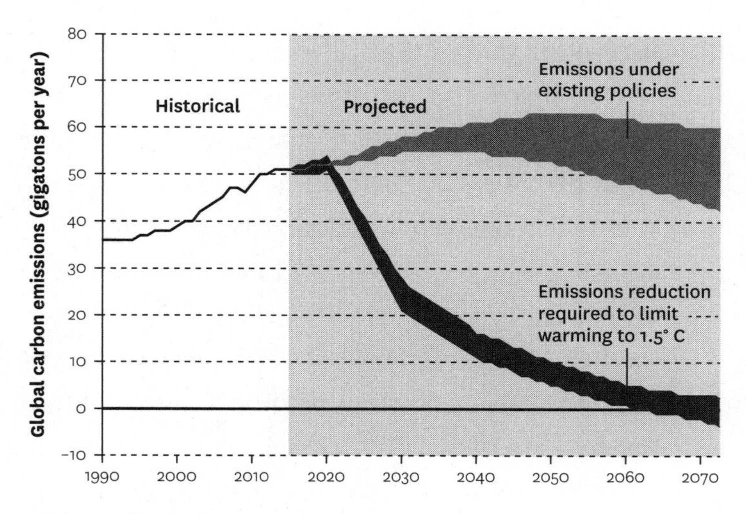

Source: Climate Action Tracker
Note: Bandwidths represent high and low emissions estimates.

If the main question for business were still "Which actions will both cut emissions and create short-term value?" we know the answer: Slash carbon in energy-intensive industries and in operations, transportation, and buildings; buy lots of renewable energy, which is strategically smart because it has been competitive with fossil fuels for years; reduce waste, particularly in critical sectors such as food and agriculture; expand the use of circular business models that minimize resource use; embed climate change metrics in corporate systems and key performance indicators; and more. Again, most companies have begun to take advantage of these "basic" opportunities and will accelerate adoption as they see the payoff grow. So let's assume that they will continue down this path. Then what?

Given the urgency, we must ask a different, and harder, question: "What are *all* the things business can possibly do with its vast resources?" What capital—financial, human, brand, and political—can companies bring to bear?

Drawing on 20 years of consulting to global corporations and working on climate change issues, I see three actions that companies must now focus on to drive deeper change:

- Using political influence to demand aggressive climate policies around the world

- Empowering suppliers, customers, and employees to drive change
- Rethinking investments and business models to eliminate waste and carbon throughout the economy

These actions may feel unnatural to some executives if they appear to put larger interests ahead of immediate shareholder profits. But the tide is turning on the very idea of shareholder primacy. The roughly 200 largest multinationals based in the United States recently declared, through the Business Roundtable, that they will no longer focus solely on shareholders or on the short run. We are at a pivotal moment as the climate crisis propels companies' growing sense of social purpose. The result, I believe, is the will needed to finally achieve this deeper change.

What's in It for Us?

Before I dig into the three areas of change, it's fair to ask why a company would commit to such challenging and possibly risky initiatives. One argument is macro/societal and the other is microeconomic. The former is straightforward: Companies need healthy people and a viable planet; with expensive runaway climate change on the horizon, they have an economic imperative and a moral responsibility to do everything they can to ensure a thriving world. As Unilever's former CEO Paul Polman says, "Business simply can't be a bystander in a system that gives it life in the first place."[4] And let's not forget that even as they pursue their own self-interest, executives sometimes just do what they believe is the right thing, which may or may not pay off—from ceasing to sell assault weapons at Dick's Sporting Goods and Walmart to funding by Apple and Microsoft of programs to reduce homelessness in their neighborhoods.

The microeconomic argument, however, is often overlooked. Stakeholders, particularly customers and employees, have increasingly high standards for the companies they buy from and work for. Business customers are demanding more sustainability performance from suppliers every year. Consumers are seeking out sustainable brands (50% of consumer packaged goods growth from 2013 to 2018 came from sustainability-marketed products), and Deloitte's global surveys show that up to 87% of the under-40 crowd—the Millennials who will make up 75% of the global workforce in five years—believes that a company's success should be measured in more than just financial terms.[5] And 9 in 10 members of Gen Z agree that companies have a responsibility to engage with environmental and social issues.

Employees are now directly pressuring their companies to do more on climate, particularly in the tech sector. In direct and public appeals, Google employees have asked their executives to cut ties to climate deniers, and Microsoft's employees staged a walkout in protest of the company's "complicity in the climate crisis." At Amazon more than 8,700 workers have signed an open letter to CEO Jeff Bezos with a list of demands, including developing a plan to get to zero emissions and eliminating donations to climate-denying legislators. Their efforts clearly played a part in pushing Bezos to announce large ambitions to be carbon neutral by 2040 and to buy 100,000 electric vehicles.

Because of pressure like this, along with increasingly dire warnings from climate scientists and global bodies including the UN, corporate efforts to reduce emissions have become table stakes—something any company *must* do to earn respect from employees and customers. And what is common and accepted practice, regardless of the short-term return on investment (ROI), can sometimes shift very quickly. Consider that nobody could prove the value of diversity and inclusion when companies first dove into that issue. Now we have good data—but the norms changed first.

I've seen firsthand how this can play out on sustainability issues. Nearly six years ago, in my book *The Big Pivot*, I advocated setting science-based emissions-reduction goals.

Virtually no companies were doing that then, and I argued with many who wondered why a company would set a goal not required by law. Now, owing to peer pressure—and because it's rational—those goals are all but standard for big companies, with about 750 signed up and more than 200 committing to 100% renewable energy. They moved from "Why would we?" to "You're a laggard if you don't."

The first companies to try the most innovative sustainability strategies are generally B Corps or purpose-driven, privately held businesses like Patagonia and IKEA, which have more leeway to experiment. The story is similar for many of the next-gen climate ideas I lay out below: Big public companies are just dipping their toes in the water, while smaller, nimbler, sustainability-focused companies take the lead. Their examples matter, because over the past decade the largest firms started emulating the midsize leaders—or just buying them. To mitigate the worst effects of climate change, more companies need to follow, and fast.

Let's return now to the three broad activities that every company, big or small, must undertake.

1. Use Political Influence for Climate Good

Given the scale of the climate crisis, business alone can't solve it. But business does have a powerful tool beyond its own practices and products: extensive and deep tendrils in the halls of political power. All over the world, but especially in market economies, companies have enormous influence over governments and politicians. Through large campaign donations and—in the United States after the Supreme Court case *Citizens United*—nearly unlimited spending on political ads, the corporate agenda gets an outsize voice in society. How can and should companies use that power?

Business's government relations have traditionally been aimed at reshaping or fighting regulations. But over the past few years many companies have, at least on the surface, been supporting some climate policy. Hundreds of multinationals with operations in the United States have signed statements such as "We Are Still In" and the recent "United for the Paris Agreement" to let the world know that they will cut emissions in keeping with the Paris Climate Accords and that they want the U.S. government to stay aboard, despite announcements that it would not. Another group of large companies called for the world to hold warming to just 1.5°C. Signatories came from every corner of the planet: Sweden (Electrolux), Japan (ASICS), India (Mahindra Group), Switzerland (Nestlé), Germany (SAP), and many other places and sectors.

But statements alone are inadequate. Companies must lobby for the policies that will lead to a low-carbon future,

FIGURE 1-2

Rising temperatures, rising risks: flooding cities

If the global temperature were to increase by . . .

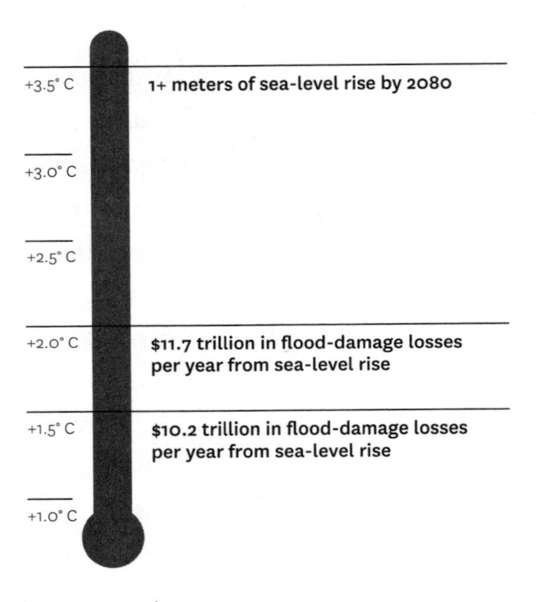

Source: World Resources Institute

and senior executives need to show up in person. Without collective government action, we have little chance of avoiding the direst outcomes of climate change. One industry—fossil fuels—has had a dominant, decades-long influence on climate policies in world capitals, and for good reason: Policies aimed at reducing emissions pose an existential threat to the business. Companies in every other sector must grasp that climate change, which may spin out of control without enlightened policies, is an existential threat to *their* businesses.

For the most part, non–fossil fuel companies engage only in occasional special lobbying days organized by the likes of Ceres, the American Sustainable Business Council, and Business Climate Leaders. Those events are important, of course, but even the groups themselves acknowledge that the number of big companies with a consistent climate-action focus is small. As Joe Britton, a former chief of staff for U.S. senator Martin Heinrich, told me, these temporary "fly-ins" are better than nothing, but they are overshadowed by the daily swarm of fossil fuel lobbyists. In response, Britton left his position to create a new lobbying organization, with the help of other Capitol Hill insiders, to deploy a fuller and more constant political message to Congress on climate.

There's also a major disconnect between what companies say about their commitments to fight climate change and what those who represent them—the trade associations or even their own government relations people—actually push for. As transparency increases, companies should worry about any gap between their sustainability commitments and their lobbying. An NGO, Australia's LobbyWatch, is calling out the mining giant BHP and others for such disconnects. And the U.K.–based influencemap.org is tracking corporate lobbying activity on climate at hundreds of companies and publicly highlighting hypocrisy.

For leaders, aggressive climate lobbying is not just about appearances; it can create advantage. If 100% of your energy comes from renewables, a price on carbon won't affect your own cost structure much. And if you make products or provide services that help reduce emissions, you benefit from tighter carbon controls. That's surely one reason that Germany's Siemens, with a portfolio of products that improve energy efficiency, states that its top political engagement goal is "combating climate change."

Hugh Welsh, the president for North America at DSM, a large Dutch company that offers nutrition, health, and sustainable-living products and solutions, can attest to this. He has worked for years to bring a business voice on climate to the halls of political power. Welsh says he does this for two reasons: principles and pragmatism. About the former, he says, "Over 10 years as president, I've developed political capital. I can use that just for strategic things for the business, but I can also use that to improve the world." About the latter, he notes that DSM serves several sustainability-focused product markets, so a proactive role on sustainability and climate policy fits its strategy.

When Welsh makes the case to skeptical executives, leaders, and trade groups—such as the recalcitrant U.S. Chamber of Commerce, with which he worked for two years to flip its position on climate—he says, "If you don't evolve your position, you'll be on the wrong side of history... your partners and customers will leave in droves."

So what policies should companies advocate? To move the world to a low-carbon future, we need bold plans in a few key areas: pricing carbon and mobilizing capital to shift to low-carbon systems; rapidly raising performance standards and phasing out old technologies for big energy users like cars and buildings; and enabling transparency and efforts to reduce human suffering.

These priorities apply in most geographies, but of course policy formation and the relationship between business and government vary widely across countries. Approaches in command-and-control economies must vary from those in sprawling capitalist systems.

Policies may take years to have an effect, so these efforts must be made soon. It's time for companies to use their substantial political influence to proactively support

Climate Policies Companies Should Fight For

A long list of possible government policies could create the conditions for rapid emissions reductions. But the following are probably the most important for business to get behind. These will fix market failures, shift capital toward low-carbon investments, and set a high bar for low-carbon products.

Implement a rapidly rising price on carbon, coupled with massive shifts in subsidies from fossil fuels to clean tech and low-carbon production methods.

Create incentives for farmers to move from industrial to regenerative agriculture.

Fund increased material capture (recycling, reuse, repair) to encourage a circular economy.

Mobilize capital and R&D that pulls public and private investment into cleaner tech. For example, the Danish aviation sector has proposed a climate tax on all flights from Denmark, earmarked for a fund to research green solutions and climate-neutral fuels.

Introduce high performance standards for the big energy users, including cars, buildings, and HVAC systems. *Encourage phaseouts and phase-ins* such as by mandating low-global-warming-potential refrigerants and net-zero buildings with renewables and banning gas-guzzlers. Some countries have set a date for stopping the sale of internal combustion engines: Norway by 2025, Sweden and Denmark by 2030, and France and Sri Lanka by 2040.

Prioritize transparency through, for example, the Task Force on Climate-related Financial Disclosures, which provides guidelines for companies reporting their material risks from climate change, and product labels with carbon-footprint information, much like the calorie and nutrition counts on food labels.

Fund resources for adaptation, such as resilience planning in cities, the relocation of citizens, and retraining for those from older sectors that will rapidly decline.

laws that make high-carbon products and choices more expensive, mobilize capital toward a clean economy, support systems change, and help deal with adaptation and the human costs of shifts to clean technology.

2. Leverage Stakeholder Relationships

At the same time, companies should wield their other superpower: vast influence over value chain partners and deep connections to their customers and employees. Big consumer products companies like P&G and Unilever often rightly brag that they serve billions of people every day. More than 275 million people visit a Walmart every week. Companies employ hundreds of millions of us. And with nearly $33 trillion in revenues across the *Fortune* Global 500 alone, it's safe to assume that many trillions go to suppliers. Imagine if companies used those touch points, their buying power, and all their communications and advertising clout to catalyze change across business and society.

Suppliers

In recent years corporations have ratcheted up the pressure on their suppliers to operate more sustainably. Big buyers increasingly want to see progress—backed up by data—in a supplier's carbon footprint, resource use, human rights and labor performance, and much more. General Mills, Kellogg, IKEA, and Hewlett Packard Enterprise have all set science-based carbon goals for their suppliers. Others, including GSK, H&M, Toyota, and Schneider Electric, have committed to carbon neutrality or negativity (eliminating more carbon than is produced) in their entire value chains by 2040 or 2050.

Commitments like these are becoming the norm. But what else is possible? What are boundary-pushing companies doing to drive change? I see future supply-chain climate leadership in three key areas: providing capital, driving innovation and collaboration, and using purchasing power to choose suppliers on the basis of emissions performance.

Financial assistance and capital. Making a business more sustainable is profitable, but it may still require investments and capital. Companies that ask suppliers to change how they do business can help, especially with smaller players. For example, in mid-2018, after achieving 100% renewable energy in its own operations, Apple launched the China Clean Energy Fund, a joint pool of $300 million to help suppliers buy one gigawatt of renewable energy, and the fund's first big wind farms went up

FIGURE 1-3

Rising temperatures, rising risks: food shortages

If the global temperature were to increase by . . .

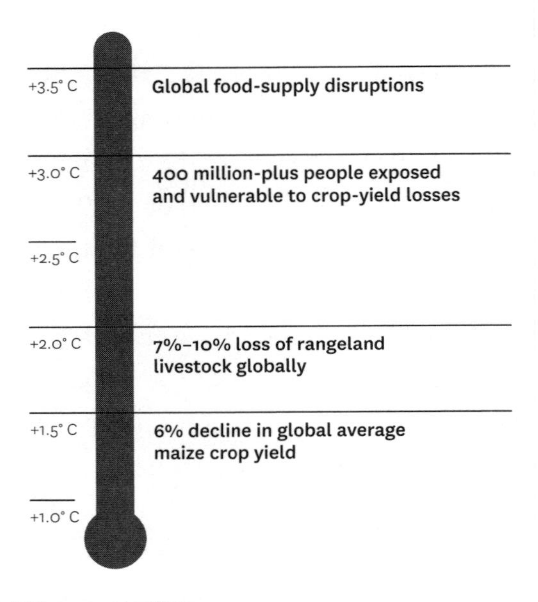

Source: World Resources Institute

the following year. Similarly, IKEA recently committed €100 million to help first-tier suppliers make the shift. In another innovative approach, an industrial company I work with, Ingersoll Rand (better known by its brands Thermo King and Trane), financed a large renewable energy project and then invited suppliers to offset their emissions by buying portions of the energy production. And beyond encouraging renewables, some leaders, such as Levi's and Walmart, have worked with HSBC and other banks to provide lower interest rates to suppliers that score well on sustainability performance.

Joint innovation. I also recently watched the head of procurement at Ingersoll Rand tell hundreds of suppliers that his company would no longer choose vendors on the basis of pricing and quality alone. Now, he said, suppliers would need to innovate *with* the company to make its products more energy and carbon efficient. This is a great way to drive value chain innovation, but sectorwide collaboration can have an even bigger impact.

Consider that Walmart and Target, which are traditionally competitors, worked together with the NGO Forum for the Future (on whose board I serve) to create the Beauty and Personal Care Sustainability Project—a creative attempt at improving the environmental and social footprint of all the products we put on our bodies. They brought together big CPG companies such as P&G and Unilever and their chemical suppliers to rethink ingredients, packaging, and more to reduce health and environmental impacts. Apple has dived deep into its supply chain to make its ubiquitous tech products lower carbon, including through a joint venture with Rio Tinto and Alcoa to develop and commercialize an aluminum-smelting process with vastly lower greenhouse gas emissions and lower costs.

Purchasing power. For years many companies have agreed to work with lagging suppliers to improve their sustainability performance. But the world can no longer afford to wait for slow adopters. Companies should cut them loose and shift their purchasing dollars toward the low-carbon leaders—which are often the best-run suppliers anyway. VF Corporation, the home of brands such as Vans and the North Face, stopped buying leather from Brazil because government policy there was encouraging Amazon rain forest destruction.

Retailers should make carbon performance a buying priority. Mainstream mega-retailers like Walmart and Target have pressured suppliers for years to make their offerings more sustainable, but they could do much more to support those that are best at reducing emissions in their operations or through their products. They could, for example, permanently (not just on Earth Day) devote endcaps or special promotion areas—their highest-value real estate—to drive business to the lowest carbon-emitting suppliers while satisfying growing customer demand for green products. It's a win-win, but it's not normal practice yet.

Customers

The core thing companies are doing—and must continue to do—is helping customers reduce carbon emissions by developing and offering products that produce fewer emissions throughout their life cycles. We're seeing great innovation, and customer buy-in, for lower-footprint products in the biggest carbon-emitting sectors: electric vehicles in transportation; efficient heating, cooling, and lighting in buildings; and tasty alternative proteins in food and agriculture.

Manufacturers and retailers are also working to increase the use of recycled materials and reduce the amount of material used in packaging—all the way to zero in some cases. A group of British retailers, for example, has teamed up to change how some products leave the store. Consumers can fill their own bags and jars from bins of dry goods (grains, beans, nuts, and so on), laundry detergent, and shampoo. Some commercial products are trying to

FIGURE 1-4

Rising temperatures, rising risks: nature's collapse

If the global temperature were to increase by . . .

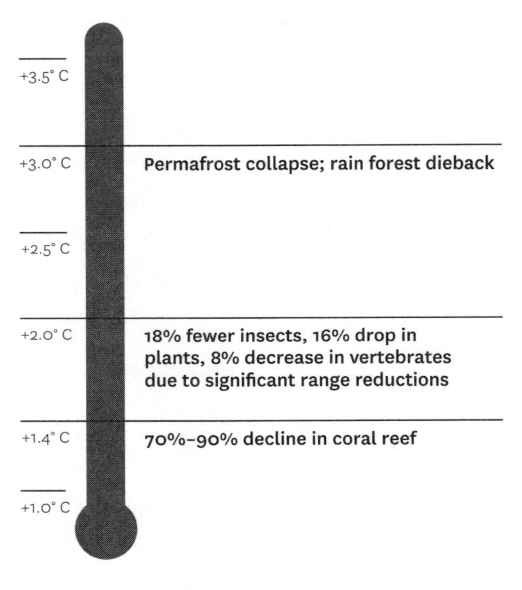

Source: World Resources Institute

go even further: After making each tile of its prototype carbon-negative flooring, Interface explains, "there is less carbon dioxide in the atmosphere than if it had not been manufactured in the first place."[6]

But businesses need to make products like these mainstream and then go beyond the direct impacts of their products on customers to drive deeper change. Here are three possible ways forward:

Help customers use less and mobilize. The two most aggressive actions companies can take with consumers are encouraging them to reduce consumption and engaging them in climate activism. Zurich-based Freitag, which makes bags from recycled materials, lets customers create a new look by switching bags with other customers. And Patagonia (always a radical company) is teaching its customers how to repair its clothes so that they don't need to buy new items. These companies may risk selling less, but they're building trusted brands with a loyal following. And discouraging consumption hasn't hurt Patagonia in the least: Sales have quadrupled over the past decade, reaching an estimated $1 billion. Going further, the company is using the trust it has built to mobilize consumers, through its Patagonia Action Works initiative, to engage with grassroots environmental groups in Europe and the United States.

Use communications to educate and inspire consumers. Companies can make more effective use of two channels in driving climate discussions: packaging and advertising. How? The Swedish oat drink brand Oatly, for example, reports product carbon emissions on its packages and points consumers to information on the climate benefits of eating plant-based products. Ben & Jerry's used the packaging and launch of an ice cream flavor, Save Our Swirled, to raise awareness about the Paris Climate Accords in 2015. IKEA surveyed more than 14,000 customers in 14 countries to understand their attitudes and how best to motivate climate action through advertising; the resulting framework is designed to guide its communications.[7] In the fall of 2019 the household products company Seventh Generation donated advertising airtime on the *Today* show to help promote the Youth Climate Movement.

A new collaborative initiative seeks to make promotional activities like these the norm. Launched recently by Sustainable Brands (on whose advisory board I sit)—along with some big names such as PepsiCo, Nestlé Waters, P&G, SC Johnson, and Visa—the Brands for Good program commits participants to encourage sustainable living through their marketing and communications and, even more ambitious, to transform the field of marketing to support that goal.

Choose business customers wisely. The efforts described above focus on traditional consumers. But companies need to direct equal attention to their business customers. As with suppliers, they must stop enabling customers that are either not addressing climate change or, more to the point, part of the high-carbon economy. Banks, venture capital and private equity funds, consulting companies, legal firms, and other service providers should ask tough questions about whom they're supporting. Helping companies be "better" at extracting or burning carbon-based fuels is actively moving the world in the wrong direction, and it dwarfs any carbon reduction a service business pursues in its own operations.

In the investment world, a movement to divest from fossil fuels is taking off, spearheaded by a group of investors with $11 trillion in assets. Norway's $1 trillion sovereign wealth fund is likewise dumping investments in many oil and gas companies.

Other service companies, such as consulting giants and law firms, that still work with carbon-intensive industries should be helping them make the permanent pivot necessary to survive. That means helping fossil fuel companies sunset their core business over the next few decades and completely shift their portfolios and business models toward clean options. Tech companies have to do some hard thinking as well. One of the reasons

Amazon's employees rebelled was the company's announcement that its cloud business would help oil and gas companies accelerate exploration. Stakeholders will continue to ask probing questions about what companies stand for and whom they support—and companies will have to have an answer.

Employees

In the battle for talent, especially for Millennials and Gen Z, companies must prove that they are good citizens. Surveys consistently show that people under 40 want to work for employers that share their values. As Unilever's sustainable living plan gained steam in the mid-2010s, the company became the most sought-after employer in its sector. Top executives I've worked with at Unilever cite its sustainability leadership as key in attracting and retaining talent. The benefit flows both ways: Companies need their employees' commitment and buy-in to achieve their sustainability goals.

To reinforce this relationship, companies must build sustainability and climate action into their regular incentive structures and systems—that is, pay everyone from the C-suite on down to cut carbon. They are secretive about the exact percentages, but the most commit-

FIGURE 1-5

Rising temperatures, rising risks: heat waves

If the global temperature were to increase by . . .

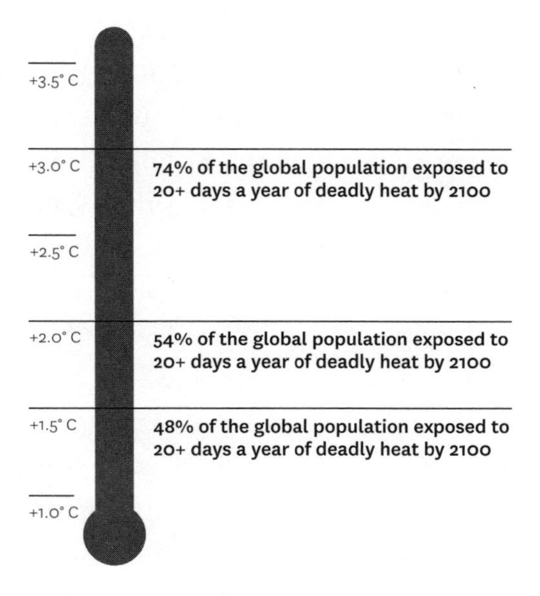

Source: World Resources Institute
Note: According to research published in *Nature Climate Change*, "deadly heat" is the threshold beyond which air temperatures, humidity, and other factors can be lethal.

ted companies I've seen tie at least a quarter of bonuses to sustainability key performance indicators (KPIs). It's time to increase that.

Can companies go even further and proactively support their employees' values by helping them drive change in the world around them? Some organizations already do. During the 2018 U.S. election, more than 100 of them, including Walmart, Levi Strauss, The Gap, Southwest Airlines, Kaiser Permanente, and Lyft, joined the Time to Vote initiative, giving employees time off to be good citizens. Some even encourage direct climate activism. Having identified the "climate emergency" as a top employee concern, the $1 billion cosmetics retailer Lush closed 200 shops in the United States to allow employees to join global climate marches last September. A Lush representative told me that during Canadian marches the company also shuttered 50 shops and offices for 20 manufacturing and support teams.

Atlassian, the fast-growing Australian enterprise software company with a $30 billion market cap, also encourages employees to become climate activists. As the company's cofounder Mike Cannon-Brookes wrote in his blunt blog, "Don't @#$% the planet," Atlassian gives employees a week each year to volunteer for charity, and they can now use the time to join marches and strikes. He wants them to "go further and volunteer their time to other not-for-profit groups with a focus on climate."

Employees want to work for a company that stands for something. But they increasingly also want the freedom to express what *they* stand for. So ask them what they care about—especially younger and newer employees—and help them live their values.

3. Rethinking the Business

Flexing political muscle and reconceiving stakeholder relationships must happen quickly. But it is also time to think big, to look for new possibilities, and to question core assumptions about consumption and growth in the economy—that is, to go far beyond simply slashing energy use and buying renewables. Today the possibilities are broad, with everything from reducing food waste to developing circular business models falling under the umbrella of "climate strategy." Now is the right time to think critically and creatively about how *all* products and services in every sector are created and used and to squeeze carbon out of every step in the value chain. Some of this is tactical—for example, working with suppliers or customers to reduce their emissions, as discussed. But at the strategic level it can mean rethinking the company's investments and business models entirely. Here are some ways to do just that, focused on two key areas.

Risk and investments

Companies deploy capital and make investment decisions in multiple ways. With some important changes in how they think about financing and investment, much more capital could flow to low-carbon activities.

Consider the idea of return on investment. In most companies, to get internal funding, a project must achieve a predetermined rate of return (or hurdle rate) that will pay off relatively quickly. This approach to ROI is flawed. It generally measures the "R" in straight cash, without allowing for more strategic or intangible value. It's also agnostic as to whether the investment moves the company down a more sustainable path. We need to use this tool differently to shift to low-carbon investment choices.

Smart tweaks to two internal processes—capital expenditures and hurdle rates—can do a lot of good. J. M. Huber, a family-owned business that manufactures nature-based ingredients for the food and personal care industries along with components in home building, developed a more holistic approach to optimizing capital deployment. The chief sustainability officer and the CFO worked together to shift the capex process to factor in intangible benefits such as community engagement, customer perceptions, employee attraction and retention, and business resiliency (for example, solar array projects that insulate the business from fossil fuel energy price shocks).

Companies should set their hurdle rates more strategically and allow some investments more leeway, with a strong bias toward funding carbon-reducing projects. If, for example, constructing an energy-efficient building—one that will save money and carbon over its lifetime—costs more up front or requires more than a few years to pay off, isn't it still a smart investment on a 40-year asset?

Another wise investment shift involves levying an internal carbon price on companies' own operations to encourage emissions reduction. More than 1,400 organizations now use internal pricing in some way, but the norm is to use "shadow" prices with no money changing hands. That approach isn't strong enough. Early leaders like Microsoft, Disney, and LVMH have been collecting *real* money from divisions or functions related to their emissions. That "tax" revenue is reinvested in energy efficiency, renewables, or offset projects such as tree planting. All companies should use this strategy to help fund low-carbon projects and to prepare the business as government-imposed carbon taxes become more common.

FIGURE 1-6

Rising temperatures, rising risks: water uncertainty

If the global temperature were to increase by . . .

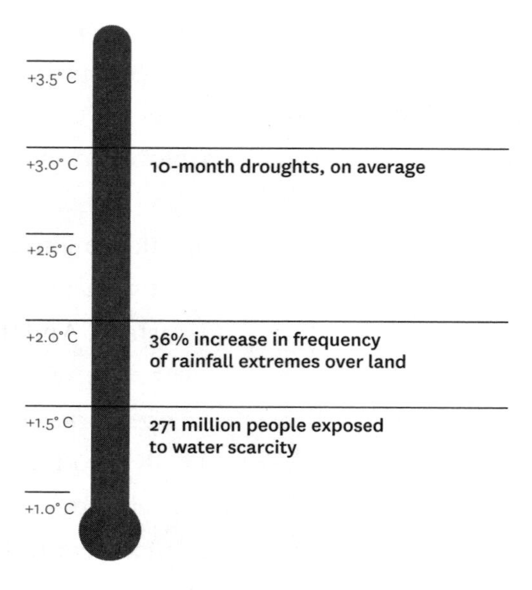

Source: World Resources Institute
Note: According to the NOAA, "extreme rainfall" can be loosely defined as a month's worth of rain for a given region falling in a single day.

A more recent strategy is to use financing tools such as green bonds, now a $200 billion market, in which the proceeds from bond purchases go to environmental and climate projects. The Italian energy group ENEL is trying something a bit different, issuing a bond tied to a KPI measuring the company's performance against the UN's Sustainable Development Goals. If ENEL misses its target of increasing renewable energy to 55% of its installed capacity, it will pay 25 basis points more to bondholders. Although the funds raised are not tied to a specific use, as they are with conventional green bonds, the instrument clearly supports emissions reduction.

Perhaps the biggest move a company can make is to rethink where to place its R&D bets. In a telling seismic shift, Daimler announced that it would no longer invest in research on internal combustion engines and would put billions toward electric vehicles instead. And the CEO of Nestlé, Mark Schneider, spoke recently about investing in plant-based proteins, which have a *much* smaller carbon footprint than conventionally produced meat, saying, "A Swiss franc we spend developing the burger is a burden to this quarter's profits. Next year or the year after, it will come back to us if we do our job right."[8] Seeing returns on a fast-growing new market within a year or two sounds like a good deal.

New business models

The level of carbon reduction that the Intergovernmental Panel on Climate Change says is required to head off catastrophic warming—cutting emissions in half by 2030 and to zero by 2050—is daunting. Everything discussed here will move us much more quickly, but some fundamental changes are needed in how we think about products, services, and consumption. Current business models and delivery methods can lock us into more material- and energy-intensive pathways. And some sectors, the most carbon intensive, will need to exit core businesses.

Consider Philips Lighting, which launched a "light as a service" model, through which business customers pay Philips to install and manage their lighting rather than purchase a lighting system themselves. This flips Philips's traditional model on its head: Instead of trying to sell as many bulbs as possible, under this program, the company manages the provision of light as frugally as it can, using longer-lasting, more-efficient products that slash material and energy use. In a larger-scale transformation, the energy company Ørsted—formerly known as Danish Oil & Natural Gas—anticipated the decarbonization of the global economy and began pivoting from its core business a decade ago. It has since sold off most of its fossil fuel assets and has become the world's largest builder of offshore wind farms. And just a few years ago, the idea that meat-based McDonald's and Burger King would both be selling plant-based "burgers" seemed far-fetched. But they, like Ørsted, may be thinking strategically about what the coming low-carbon economy means for their business.

The Next Level of Action

There's no doubt that companies are doing a lot on climate, including cutting emissions and setting aggressive carbon goals for operations, supply chains, and their innovation agendas. But it's not enough. The science is getting away from us, and we're losing the relatively stable planetary temperature range that allowed us to build our society over the past 10,000 years. Companies have many levers to pull to truly change business as usual, but most remain stuck in old thinking. Climate action is usually focused on incremental change. And even when they're setting a big goal like going to all-renewable energy, companies have waited until every project makes money quickly. Now they need to mobilize *all* corporate assets, hard *and* soft, to tackle this shared, unprecedented problem at the scale it requires.

Next-gen climate actions, as they become an expected part of business, will create significant long-term value.

They will help companies build closer, lasting connections with key stakeholders; create clear and consistent regulatory environments that enable more sustainable practices that lower costs; and drive deeper, more-disruptive (or what I call *heretical*) innovation. Throw in the substantial intangible value—employee attraction and loyalty, lowered risk in supply chain, resilience, license to operate, societal relevance, and preparation for a very different future—and you have a powerful business case.

But it's also well past time to recognize that aggressive climate action is necessary if humanity is to survive and thrive. Business and society won't succeed unless and until we do all we can to tackle climate change.

Companies need to mobilize in three ways to deal with the unprecedented global problem of climate change:

✓ **Use political influence to demand aggressive climate policies.** Proactively support laws that make high-carbon products and choices more expensive, mobilize capital toward a clean economy, and help deal with adaptation and the human costs of shifts to clean technology, including resilience planning in cities, relocation of citizens, and retraining workers.

✓ **Empower key stakeholders to drive change.** Limit your suppliers to only those who have low emissions or work with them to create more energy- and carbon-efficient products. Offer your customers products that produce fewer emissions throughout their life cycles and provide ways for them to use less, thereby reducing their carbon footprints. Build sustainability and climate action into incentive structures for your employees by paying everyone (from the C-suite on down) to cut carbon.

✓ **Rethink investments and business models to eliminate waste and carbon.** Think about how all products and services in every sector of your company are created and used, and squeeze carbon out of every step in the value chain. Then, rethink where your company is taking risks and consider new business models.

NOTES

1. Christian Mumenthaler, "Swiss Re Retains Its Strength in a Challenging Year and Commits to a More Sustainable Future," Swiss Re, February 21, 2019, https://reports.swissre.com/2018/business-report/the-year-in-review/statement-from-the-group-ceo.html.

2. Mattha Busby, "Capitalism Is Part of Solution to Climate Crisis, Says Mark Carney," *The Guardian*, July 31, 2019, https://www.theguardian.com/business/2019/jul/31/capitalism-is-part-of-solution-to-climate-crisis-says-mark-carney.

3. UN Environment Programme, "Emissions Gap Report 2019," November 26, 2019, https://www.unenvironment.org/resources/emissions-gap-report-2019.

4. Paul Polman, "Business, Society, and the Future of Capitalism," *McKinsey Quarterly*, May 2014, https://www.mckinsey.com/business-functions/sustainability/our-insights/business-society-and-the-future-of-capitalism.

5. Tensie Whelan and Randi Kronthal-Sacco, "Research: Actually, Consumers Do Buy Sustainable Products," hbr.org, June 19, 2019, https://hbr.org/2019/06/research-actually-consumers-do-buy-sustainable-products; Deloitte, "Millennials Want Business to Shift Its Purpose," The Deloitte Millennial Survey 2016, https://www2.deloitte.com/us/en/pages/about-deloitte/articles/millennials-shifting-business-purpose.html.

6. Interface, "Interface Unveils Prototype Carpet Tile to Inspire New Approaches to Address Climate Change," June 8, 2017, https://www.interface.com/US/en-US/about/press-room/carbon-storing-tile-release-en_US.

7. IKEA, "Climate Action Starts at Home," IKEA Climate Action Report, 2018, https://www.ikea.com/ms/en_US/pdf/reports

-downloads/IKEA%20Climate%20Action%20Report%20 20180906%20(002).pdf.

8. Jack Ewing, "Nestlé Says It Can Be Virtuous and Profitable. Is That Even Possible?" *New York Times*, November 15, 2019, https://www.nytimes.com/2019/11/15/business/nestle-environment-sustainability.html.

Adapted from content posted on hbr.org, January 2020 (product #BG2001).

WHEN ENVIRONMENTAL REGULATIONS ARE TIGHTER AT HOME, COMPANIES EMIT MORE ABROAD

by Itzhak (Zahi) Ben-David, Stefanie Kleimeier, and Michael Viehs

If we want to avoid the most damaging effects of climate change, countries need to work together to limit pollution and keep global warming to under 1.5°C. But countries differ in their approaches toward environmental regulation. Some are trying to reduce carbon dioxide (CO_2) emissions by enforcing strict environmental policies. Others are considering withdrawing from the Paris Agreement designed to collectively combat climate change—or already have.

This gap makes a difference. Our research finds that global emission levels are *lower* for countries with tighter domestic environmental regulations.[1] One reason for this, we find, is that companies, which are the biggest contributors of CO_2 emissions, emit less CO_2 at home when domestic environmental regulations are strict. However, these companies also emit more abroad, particularly in countries with laxer environmental standards.

Fortunately, the higher foreign emission levels do not outweigh the reduction at home. But if we want to progress, countries will have to take collective action to bring down overall global emission levels further.

Pollution Havens

Academics have long argued that there is a symbiotic relationship between countries' environmental policies and the degree to which companies pollute. But the causality between these two factors could be bidirectional: Countries may adopt lenient environmental policies to attract foreign companies, and/or companies may transfer polluting operations to those countries. Either way, the idea is that polluting activities are more likely to be performed in countries with loose environmental policies. Economists call this hypothesis the *pollution haven hypothesis* (PHH).[2]

Until now, this hypothesis was typically tested using data at the country or industry level, mostly without access to direct measurement of pollution. For example, several studies correlate aggregate industrial activity (for example, FDI as a proxy for pollution) and the stringency of environmental laws in home countries compared to foreign countries. And studies that use firm-level data do not observe pollution activities directly and often infer them from other variables (such as a firm's location or manufacturing decisions).

We used a unique data set of CO_2 emissions to see whether companies actually pollute more in countries with weak environmental laws and enforcement. Our data includes information for over 1,800 international firms and reports their CO_2 emissions for each country in which the companies operated between the years 2008 and 2016. Although the emission figures are self-reported, they are frequently audited and used by institutional investors who monitor the firms. We combined these data with information on the stringency of national environmental policies from the World Economic Forum (WEF).

We conducted various tests to investigate the implications of the pollution haven hypothesis. In particular, we looked at whether companies tend to pollute less at home but more abroad, given the strictness of environmental policies in their home country. Also, we explored the effects of the "regulatory distance" between a firm's home country and each foreign country (in other words, the difference between the strictness of the environmental regulation in the home and the foreign country), on the likelihood of that firm polluting in that foreign country.

We documented a number of important correlations. Our findings reveal that in countries with tight environmental regulation, companies have 29% lower domestic emissions on average. On the other hand, such a tightening in regulation results in 43% higher emissions abroad. Importantly, although companies appear to emit more CO_2 abroad, stricter environmental policies at home are associated with lower global pollution overall. Tightening of environmental policies in the home country (a one standard deviation increase in environmental regulation within the sample of investigated countries and companies) is associated with about 15% lower global CO_2 emissions overall.

We also explored different factors that might make companies more or less likely to pollute abroad. For companies that are considered to have good governance (using a standard governance scoring that is often used in the literature), we found that the observed effects are generally weaker: When the home country sets strict environmental policies, companies commonly considered to have good governance structures produce fewer emissions at home and export fewer emissions to foreign countries. Thus, it is predominantly the firms with weak governance structures that behave according to the PHH and conduct their polluting activities abroad.

This is interesting, as companies can face a trade-off between pollution and value. At least in the short run, companies may prefer to pollute in order to save on the costs associated with clean production.[3] But good governance mechanisms, such as strong shareholder monitoring, may dissuade managers from pursuing such short-term goals and push them toward production with lower emissions. Good governance is generally associated with an investor base that values corporate responsibility practices and puts pressure on management to pursue socially and environmentally responsible goals, including lower emissions.

We also found that some industries are more susceptible to exporting pollution abroad than others. The heaviest polluters are firms in industries such as electricity, gas and refineries, steam and air-conditioning supply, air and water transport, and mineral and metals producers. Our study shows that companies in these industries do not reduce emissions at home, while at the same time they export more pollution abroad. We argue that this is because strict environmental policies are costly for companies in high-polluting industries, causing them to attempt to mitigate some of the cost by exporting pollution to foreign countries. Consequently, policy makers might have a greater impact on global emissions if they target these high-polluting industries.

Where do companies "export" their pollution? By drawing on empirical methods used in the international trade literature, we found that companies pollute more in foreign countries with the least regulation, whose standards fall far below those of their home countries. For example, Trinidad and Tobago, Bosnia and Herzegovina, Slovakia, Suriname, and Barbados are frequently among the top five countries that annually receive most exports of direct CO_2 emissions relative to their GDP. They are also among the countries with the weakest environmental regulation.

Overall, we find support for the idea that companies perform their polluting activities outside their home country when domestic environmental policies are becoming relatively stricter than abroad. Strict environmental regulation is associated with lower company-level emissions at home, but companies then seek to continue those polluting activities elsewhere. Fortunately, we also find that companies pollute less on a global level when their home countries impose strict environmental policies.

Our findings suggest that national regulation can be beneficial, but it can only do so much to effectively combat pollution and climate change. Because of the potential for regulatory arbitrage, countries need to take concerted action to ensure that the overall CO_2 balance will not increase. If no coordinated effort is undertaken to address climate change, major stakeholders, such as large companies, will find ways to at least partially circumvent strict environmental regulations and move their CO_2-intensive activities elsewhere. Future research should explore the effects of whether changing regulation could alter firm behavior locally and in foreign countries.

To avoid the most damaging effects of climate change, countries need to work together to limit pollution and keep global warming to under 1.5°C. National regulation should help push companies toward this goal, but without global support, companies are still prone to polluting abroad.

- ✓ Research finds that in countries with tight environmental regulation, companies have 29% lower domestic CO_2 emissions on average, but they also have 43% *higher* emissions abroad.
- ✓ Polluting activities are more likely to be performed in countries with more relaxed environmental policies—areas economists refer to as *pollution havens*. Countries with the weakest regulations include Trinidad and Tobago, Bosnia and Herzegovina, Slovakia, Suriname, and Barbados.
- ✓ Industries most susceptible to polluting abroad are electricity, gas and refineries, steam and air conditioning supply, air and water transport, and mineral and metal producers.
- ✓ While domestic regulations are associated with about 15% lower global CO_2 emissions overall, countries still need to take concerted action to ensure that the overall CO_2 balance will not increase.

NOTES

1. Itzhak Ben-David, Yeejin Jang, Stefanie Kleimeier, and Michael Viehs, "Exporting Pollution: Where Do Multinational Firms Emit CO_2?" Fisher College of Business Working Paper No. 2018-03-20, September 2018.

2. Gunnar S. Eskeland and Ann E. Harrison, "Moving to Greener Pastures? Multinationals and the Pollution Haven Hypothesis," *Journal of Development Economics* 70, no. 1 (February 2003): 1–23.

3. Roy Shapira and Luigi Zingales, "Is Pollution Value-Maximizing? The DuPont Case," NBER Working Paper No. 23866, September 2017.

Adapted from "Research: When Environmental Regulations Are Tighter at Home, Companies Emit More Abroad" on hbr.org, February 4, 2019 (product #H04RSG).

3

THE STATE OF SOCIALLY RESPONSIBLE INVESTING

by Adam Connaker and Saadia Madsbjerg

In 2007, the European Investment Bank issued its first green bond, a EUR 600-million equity index-linked security, whose proceeds were used to fund renewable energy and energy-efficiency projects. A year later, the World Bank followed suit, and by 2017, over $155 billion worth of public and corporate green bonds had been issued, paving the way for the Seychelles government to issue the first ever "blue bond" last year—a $15-million bond to fund marine protection and sustainable fisheries.

The success of these instruments reflects the fact that investors are increasingly conscious of the social and environmental consequences of the decisions that governments and companies make. They can be quick to punish companies for child labor practices, human rights abuses, negative environmental impact, poor governance, and a lack of gender equality. Pair this with an increase in regulatory drivers post-2008 crisis and a deepening understanding of the impacts of climate change and associated risk to performance and we begin to see more clearly the need for investment models that will better address investors' concerns.

The result has been an increasing demand for integrating environmental, social, and governance (ESG) criteria into investment decisions. In the beginning of 2018, $11.6 trillion of all professionally managed assets—$1 of every $4 invested in the United States—were under ESG investment strategies, a sharp increase from 2010, when the amount was close to just $3 trillion overall.[1]

Inevitably, the financial services sector has responded with a host of innovative financial instruments, some like those mentioned above, others quite different. The through-line that ties together these new investing models and strategies is quite simple: While they have generated competitive returns, it so happens that they all positively benefit society as well. Essentially what investors want is the performance promise of financial engineering combined with the assurance of a better tomorrow.

Many of the innovations have been driven by a collaboration between public, private, and philanthropic institutions. At the Rockefeller Foundation, we recognize the value of engaging private capital markets for societal good and have stepped in to fund the research and development of new instruments that can bring capital to cause. We have increasingly seen, firsthand, how readily these instruments meet not just investor needs but also values, and how interrelated the two can be.

Let's look at some particularly interesting examples.

Risk-Sharing Impact Bonds

Fixed income is one of the largest asset classes as determined by asset owner allocation and market size. Compared to the other asset classes, it has the lowest expected returns, hence also the lowest cost of capital. At $4 trillion, the U.S. municipal bond market is one of the largest fixed income markets globally.

Climate change is becoming increasingly important for the U.S. municipal finance sector. On the one hand,

U.S. cities need to raise more capital to implement environmental projects—many of them based on innovative climate solutions—to protect their economies and communities from the effects of climate changes (for example, green infrastructure to manage flooding and waste-to-energy microgrids to prevent power outages during hurricanes). On the other hand, if they do not show meaningful results, they won't only risk economic losses from disasters but also risk seeing an increase in their overall cost of borrowing. Rating companies such as Moody's are increasingly assessing climate risk as a negative factor when assessing credit ratings.

Environmental impact bonds—in many ways an extension of the green bonds market—offer a solution to this problem, because they can draw in investors interested in taking on the environmental risk in exchange for potential monetary reward. These securities are municipal bonds that transfer a portion of the risk involved with implementing climate adaptation or mitigation projects from the public agency to the bondholder. A good example is a $25 million bond issued by the Municipal Water Board in Washington, D.C., in 2016.

The water board used the bond to fund the construction of green infrastructure to manage storm water runoff and improve water quality. The return to investors is linked to the performance of the funded infrastructure, which allows D.C. Water to hedge a portion of the risk associated with both constructing green infrastructure and, once it's in place, how well it works. Another such bond is currently under development by the city of Atlanta, for approximately $13 million worth of green infrastructure projects in flood-prone neighborhoods on the city's west side.

In the case of the D.C. Water bond, investors receive a standard 3.43% semiannual coupon payment throughout the term of the tax-exempt bond. Toward the end of a five-year term—at the mandatory tender date—the reduction in stormwater runoff resulting from the green infrastructure is used to calculate and assign an additional payment. If the results are strong (defined in three tiers; tier 1 being best performance), the investors receive an additional payment ($3.3 million), bringing their interest rate effectively to 5.8%. If the results are as expected, there is no additional payment. And if the infrastructure underperforms, the investors owe a payment to D.C. Water ($3.3 million), bringing the interest rate to 0.8%.

Financially Passive, Socially Active Funds

One of the most interesting of the new generation of ESG–driven financial innovations can be seen in the exchange-traded fund (ETF) sector, where we are starting to see the passive investment movement linked to activism on key social and environmental issues through ESG–themed ETFs. The first ESG ETF, iShares MSCI USA ESG Select ETF, was launched in 2005, and the model has caught on so quickly that today there are at least $11 billion in assets under management across 120 ESG funds globally; stateside, the growth of assets in ESG funds is up over 200% from the past decade. BlackRock recently predicted that the investment in ESG funds will rise to more than $400 billion over the next 10 years.[2]

A good example is the NAACP Minority Empowerment ETF, or NACP. Issued by Impact Shares, a nonprofit fund manager, the specific criteria for the index, which includes companies such as Microsoft, Pepsi, and Verizon, are identified and compiled by the NAACP. The criteria assess the levels of social activism, equal opportunity, and diversity of workplaces within each company. These criteria are measured and tracked by the fund's ESG research provider Sustainalytics. The index is then constructed from the top 200 scoring companies by Morningstar, which uses a weighting methodology that maximizes exposure to companies with high scores on the NAACP criteria, while maintaining risks and returns similar to the Morningstar U.S. Large-Mid Cap Index.

Essentially the ETF allows investors to allocate their capital passively, while the NAACP takes on the role of the activist organization that directly engages with the companies on how to adopt and maintain strong practices to the benefit of investors. The management fee is 76 basis points, of which 15–25 points are used to cover expenses. The remainder goes to the NAACP in return for its engagement with the companies indexed, representing $5 to $6 a year for every $1,000 invested in the fund.

Impact Securitization

Pooling various types of contractual debt—such as mortgages or loans—and selling the related cash flows to third-party investors as securities has long been recognized as a useful way to create liquid secondary markets. Yes, the reckless use of securitization contributed to a meltdown of the financial markets. But used responsibly, securitization helps (and has helped for decades) hardworking people get affordable mortgages at rates only modestly higher than those charged to the U.S. government.

With the right incentives and financial structure in place, securitization can also be a highly effective means for gathering large amounts of (cheaper) capital in a relatively short period of time for environmental and social investments. If we can learn from the mistakes of the crisis and address their weaknesses, these tools can have a transformative role in many socially important initiatives.

The U.S. student loan start-up Sixup represents a step in this direction. Sixup is an education finance platform that gives high-achieving low-income students, including first-generation and minorities—collectively termed "Future-Prime"—a pathway to attending four-year colleges and universities. They have built a new model for assessing creditworthiness of "thin-file" students who gain admission to high-value four-year colleges but cannot afford to attend as they are shut out of the financing market due to lack of FICO scores or their parents' low-income status.

Along with loans, Sixup provides students with tutoring, job matching, and other counseling. Sixup currently counts Goldman Sachs as its largest lender. Once it has reached $100 million in total lending assets, it will test the market with securitization—a critical milestone toward scale. Over time, as their lending assets grow, Sixup plans to tap into the broader fixed-income markets, as well as more traditional securitizations. If successful, it has the potential to mobilize more than $1 billion toward the

Future-Prime market providing thousands with a stronger pathway to economic mobility.

A similarly promising securitization project is under way at MIT's Laboratory for Financial Engineering, which is developing a new investment model for orphan disease drug development with an eye to investing in multiple drug trials simultaneously. Rare and orphan diseases—like pediatric cancers or cystic fibrosis—are diseases that affect a very small part of the population (less than 200,000 patients in the United States). Research into new therapies for these diseases struggle to raise funding as the patient sizes are small, the cost of clinical trials is high, and the probability of success for any one R&D project is low.

But if you aggregate many research projects into one portfolio, at critical scale, the fund becomes more predictable and yields a more attractive risk-adjusted return on the investment, as well as a higher likelihood of success in finding cures for these diseases. This, in turn, enables the fund to raise money by issuing research-backed obligations (RBO) bonds guaranteed by the portfolio of possible drugs and their associated intellectual property. As the MIT team explains, "What one has to do to attract the vast majority of capital in the world to the early-stage space like this is to say, look, it's true, we're going to take mostly losses, but the couple of wins are going to pay us out for those losses."

The RBOs will be structured as bonds, targeting fixed-income investors, who collectively represent a much larger pool of capital and who have traditionally not been able to invest in early stage drug development. The fund is collaborating with the Harrington Project—a $340-million initiative to support the discovery and development of new therapeutic breakthroughs with a special focus on rare and orphan diseases. Successful implementation of RBOs could result in an unprecedented amount of capital available for translational biomedical medicine to treat rare and orphan diseases. It is estimated that 50% of those affected by rare diseases are children, and of those, 30% will not live to see their fifth birthday.[3]

In one sense, nothing about what drives us to invest has changed. We invest because we are planning for the future and hoping for a better, wealthier tomorrow. What has changed, perhaps, is our sense of what constitutes wealth. The ever-innovative financial service industries have responded to these changes—as they have always done—by creating new modes and tools of investment. And with these innovations, perhaps, we can all work more effectively to make the world a healthier, safer place for our children.

Investors are increasingly concerned about the environmental, social, and governance (ESG) consequences of investment decisions. In response, organizations are partnering with financial service companies to develop innovative ways to fund positive ESG projects and companies. Such options include:

- ✓ **Impact bonds:** An extension of the green bonds market, impact bonds draw in investors interested in taking on the environmental risk in exchange for potential monetary reward. These securities are municipal bonds that transfer a portion of the risk involved with implementing climate adaptation or mitigation projects from the public agency to the bondholder.
- ✓ **Socially active index funds:** ESG–themed exchange traded funds (ETFs) link passive investment to activism. They allow investors to allocate capital passively, while an activist organization directly engages with companies on how to adopt and

maintain socially responsible practices to the benefit of investors.

- ✓ **Impact securitization:** This option gathers large amounts of cheaper capital in a relatively short period of time for environmental and social investments. When the projects associated with this capital are aggregated into a portfolio, the fund becomes more predictable and yields a more risk-adjusted return on the investment.

NOTES

1. The US SIF Foundation, "2018 Report on US Sustainable, Responsible, and Impact Investing Trends," 2018, https://www.ussif.org/files/2018%20_Trends_OnePager_Overview(2).pdf; Social Investment Forum Foundation, "Report on Socially Responsible Investing Trends in the United States," 2010, https://www.ussif.org/files/Publications/10_Trends_Exec_Summary.pdf.

2. David Ricketts, "BlackRock Predicts Sustainable ETF Assets Will Top $400bn," *Financial News*, October 23, 2018, https://www.fnlondon.com/articles/blackrock-predicts-sustainable-etf-assets-will-top-400bn-20181023.

3. Global Genes, "RARE Facts," 2019, https://globalgenes.org/rare-facts/.

Adapted from content posted on hbr.org, January 17, 2019 (product #H04QYP).

DEMATERIALIZATION AND WHAT IT MEANS FOR THE ECONOMY—AND CLIMATE CHANGE

An interview with Andrew McAfee by Curt Nickisch

The industrial era brought an unbelievable rise in human prosperity. As economies grew and standards of living climbed ever higher, forests were cleared, soil was stripped, and oceans were emptied. When the United States celebrated the first Earth Day back in 1970, people were afraid the world would soon run out of food and other resources, burning it up like flash paper, gone forever. But that seemingly unstoppable tide could be turning.

Andrew McAfee, codirector of the MIT Initiative on the Digital Economy and author of the book *More from Less: The Surprising Story of How We Learned to Prosper Using Fewer Resources—and What Happens Next*, has been studying a surprising new trend of dematerialization. Thanks to new technologies and digitization, some national economies are managing to grow at the same time they use less material. This counterintuitive trend is not happening fast enough to stop the likes of climate change, but it offers some hope: that future economic prosperity may not damage the environment as badly as before.

Can you please explain what's happening? How can advanced economies grow but use fewer materials overall? What's going on?

ANDY McAFEE: It's deeply counterintuitive. I didn't believe it the first time I heard about it. I had walked around with this unexamined assumption that as economies grow—and as populations grow—they use more materials from the earth. They use more resources. You need molecules to build an economy.

I came across this wonderful essay written by Jesse Ausubel called "The Return of Nature: How Technology Liberates the Environment." He made this point, and I thought, "That would be wonderful if it were true, but that can't possibly be right. That's not how growth works." So I went to his sources. I double-checked, and I came to the conclusion he was absolutely right.

I became extremely enthusiastic about this and wanted to understand it and try to explain it, because this is a really profound change in our relationship with the planet that we all live on. We had 200, almost 250 years of the industrial era, which was a period of amazing growth, human population, human prosperity, and economies. But, wow, the industrial era was really tough on our planet. We took more from the earth year after year. We dug mines. We chopped down forests. We polluted. We killed all of the passenger pigeons. We almost killed all the buffalo and whales.

In some ways, this was a tough chapter for the planet Earth, and it looked like there was this trade-off between our prosperity and the health of the planet. I don't believe that trade-off has to exist anymore because we're demonstrating that we can grow economies, increase the population, grow prosperity, and improve the human condition while also taking better care of the earth and treading more lightly on it.

Which is a profound idea because we all understand that business and its quest for efficiency and being more productive is on a path to using things better and more efficiently. But it still seems like it's always a plus one or a plus 0.75 endeavor, where you're creating a new product to sell. You may be using things more efficiently, but you're still creating a new product, and people are throwing out their old ones. But you're saying this is actually turning to the point where we're using less material. We're actually reversing the trend and using less iron, for instance.

Less nickel. Less gold. Less fertilizer. Less water for irrigation and less timber. Less paper. Less of just about all of the molecules. All the things that you build an economy out of.

There's an important distinction here. There are two kinds of less. There's less per capita—in other words, less timber per American. And then there's less timber year after year, by all Americans put together, which is a much more profound phenomenon. That's essentially saying that all Americans' total footprint on the planet is shrinking over time. That's what is going on and that's what I wrote *More from Less* about.

You've actually made predictions about where America's consumption of certain materials is going to be in the next decade.

Well, you could think that this absolute dematerialization is happening. It's pretty clear in the evidence. But maybe it's just a temporary lull before our voracious appetites kick in again and cause us to dig more mines, chop down more forests, do all these things, and take more from the earth. I don't think that's what's happening, and I don't think that's what's going to happen. I think the downward trend in a lot of materials is going to continue, and we've already seen some plateaus. For example, total American energy use in 2018 was only a tick—about a quarter of a percent—bigger than it was in 2007. The economy's a lot bigger than it was in 2007, yet our energy use has flatlined.

What I think is going to happen is that that total energy use is actually going to start going down, just like we saw with lots of other resources. The main reason I believe that is extremely simple. You highlighted businesses and their relentless quest for growth and profit. That's absolutely true. That's at the heart of the capitalist system.

We need to keep in mind businesses' quest for higher profits is always a simultaneous quest for lower costs. A penny saved is a penny earned. Materials cost money. What I think is going on is that in this era of amazing technologies, you have the computer, the network, the hardware, and the software. In this era of amazing technologies, we have this widespread opportunity to swap out atoms for bits. And because you don't pay typically for each additional bit, companies are taking technology up on that offer over and over and over, in big ways and small ways, and that adds up.

Let's dig into a couple of examples just so that we can really picture what's happening. Let's talk about timber. Why is the United States now using less timber than it used to?

Well, think about the things that we use timber for. We used to use timber in the 19th century for all kinds of things, including building ships and ships' masts. Now, we pretty clearly don't do that anymore. A lot less appreciated, we use immense amounts of timber to build railroads all across the country. Railroad ties were made out of wood for a long time, soaked in creosote. We don't do that anymore. We make them out of concrete.

So, because of that substitution effect, and because we just don't make ships out of wood anymore, our total use of timber is at least 20%, maybe 30%, lower than it was in 1990, which was the year of peak timber in the United States.

Simultaneously, the year of peak paper in the United States was 1990. Right now we're about 40% below the level of paper that we used in 1990. The answer there is even easier for me to see. When was the last time you printed out maps to get from point A to point B? You don't print out memos or documents nearly as much anymore because we look at screens and because we have GPS systems on our smartphones. So, you could look at the paper generated earlier in the computer age—all that 11 × 17-inch fanfold paper, all the stuff we used to print out—and think that the computer age would be bad for total paper use and for cutting down trees. That's not true at all. It's really good. It's finally letting us get past peak paper.

An even crazier thing that I came across when I was researching the book was that our total use of cardboard is basically about as big as it was in 1995. I found that extremely hard to believe because of all the Amazon packages that show up outside my door almost every day. But again, that's the very visible phenomenon. The much less visible phenomenon is the fact that Jeff Bezos realizes that I get zero value from that cardboard, so he and his colleagues are working to reduce the total amount of packaging material that you need to get something to my front door.

So, there are all these efficiencies happening back farther in the supply chain and lots of innovation to make cardboard boxes lighter, less materials intensive, again, because this stuff costs money. It's hard to believe, but all of those savings add up, and they bring us to a point where our aggregate cardboard use is where it was almost 25 years ago. It's a crazy phenomenon.

Are we hitting peak paper in other countries, and are we hitting peak paper and peak timber in developing nations that aren't in a position to muster the power of these technologies?

I came across this research that said humanity as a whole, globally, probably hit peak paper in about 2013. So, total human use of paper is finally going down. For the other materials that you build an economy out of, I don't think that's the case. However, I can't say that with 100% confidence because the data gets spotty and much, much lower quality when you move from the United States to the entire world.

Globally, I don't think humanity is at peak stuff yet because there are too many low-income countries that are rapidly growing and becoming more prosperous, and you have to build an economy. You have to urbanize and build infrastructure. That's a materials-intensive process.

But one of the wild things I learned writing this book was that if you look at satellite imagery about urbanization (as opposed to relying on countries to give you a list of their cities and how many people live in them, which can be really inconsistent), we are an urbanized species as humans. And the great majority of us already live in an urban environment.

We've built up a lot of the physical infrastructure that we're going to need. We'll need more. Nigeria is clearly going to need a lot more stuff because its population is going to grow so much in the decades ahead. But Nigeria is also not going to lay a copper telephone network like the currently rich world did in order to have telecommunications infrastructure. And I'm pretty sure Nigerians are not going to buy as many cars per capita as Americans did as they were becoming prosperous; they're not going to build as many coal plants because technology has evolved. So, I'm not saying we're globally at peak stuff, but I *am* saying we might be surprised at how quickly we get there because the countries that are becoming more prosperous today are going to follow very different technology and materials paths than the United States and other currently rich countries did.

This dematerialization trend of moving away from peak is also happening in agriculture. Do you think

the U.S., for example, is going to farm less land in the future but still produce more food?

The reason I'm so confident about that is that's exactly what we've been doing. One of the wildest graphs that I drew when I was writing the book is a graph of total U.S. crop tonnage. We're an agricultural juggernaut, and our total tonnage of crops goes up year after year. Included in that graph is total fertilizer use for all U.S. agriculture, total water use, and total land use. All of those are now going down, sometimes by quite a lot. I still find it a crazy phenomenon, but we are very clearly getting more from less of all the inputs to growing a crop, except sunlight. Fertilizer, water, and land, we're getting more crops out year after year.

This is a very, very broad phenomenon. The USGS tracks, I believe, 72 different materials. I think all, but around six of them are now on a downward trend. The biggest exception is plastics, but there's something interesting going on there as well. Overall plastics consumption used to grow even more quickly than the economy did. Plastics are incredibly useful. We use them all over the place. Now, plastics use is still increasing, but it's increasing more slowly than the overall economy is. So, we're already hitting relative dematerialization with plastics. I don't know when exactly, but we're going to hit peak plastics, even as our economy grows, and we're going to start using less of it.

Is the trend toward dematerialization fast enough? We are at a time when we see the rising temperatures and environmental changes that are happening at a worrying pace that even with this growth—even with population growth—you can see how dematerialization and these technological trends are helping. But are they helping enough? Will they move quickly enough to make a difference?

Decarbonization is not happening quickly enough. It's happening in the rich world. It's not happening in the lower-income world. And in general, it is absolutely not happening quickly enough.

Think about a bathtub. We've got a tap, we've got a faucet that puts water in the bathtub, and we have a drain that takes water out. That bathtub is the earth's atmosphere. The drain is how quickly carbon leaves the atmosphere. The faucet is how quickly we're putting it in. The problem is that drain is not operating very quickly. The carbon that we put in hangs around in the atmosphere for decades or centuries.

So, we have to shut down the faucet even more quickly than we might think and find ways to increase the speed of the draining. We're not doing it quickly enough, and my huge frustration is that we know the playbook. If we were actually interested in decarbonizing our economies, we know the playbook for doing it. We're just doing a lousy job of following it.

When you look ahead, technologies are impossible to predict, right? Do you feel like the trends are moving in a good direction, or we'll hit another plateau and this dematerialization trend will flatten out again?

Flatten out, or even reverse, right? I don't think that's what's going to happen. Capitalism and tech progress are rising around the world quicker than they ever have before. In the year 2000, there were about 12 mobile phone subscriptions for every 100 people on the planet. Less than 20 years later, there are more mobile phone subscriptions than there are human beings on the planet.

The world is interconnected for the first time with powerful devices. The smartphones that people in low-income countries have are about as powerful as the first smartphones that you and I had. So, technology is spreading around the world really quickly, and that means that the people who have those devices are not going to buy a number of other things. They're not going to buy film cameras, camcorders, alarm clocks, or answering machines. That is a big heavy pile of stuff that we used to have to exploit the earth to generate. That's not what's going to happen in the future. Those people are going to be getting transportation options via their phones. The technology means that there are logistic networks. Their trucks, planes, and whatnot are not going to drive around mostly empty most of the time. They're going to have very high yield and high efficiency.

I want to be clear. I am not a utopian. We have real challenges ahead of us, but I think that this particular challenge of dematerializing our growth and our consumption, I'm really confident this is going to continue.

I have a 22-month-old, and I think a lot about the world that I grew up in and the world that she's going to live in. You're at this interesting place because you're an economist, you study the digital economy, and you're thinking about the future. What would you recommend to somebody who is just setting out on their career or an education, and they are looking at all the different places where they could live a meaningful life and make a difference in this global challenge? Whether it's government and policy or whether it's in technology and science, where do you feel we need more people, and what would you recommend to somebody who is trying to figure out the path that they can make a big difference in?

I get this question all the time. I feel like a big part of the reason I get it from parents of children of all ages is that we are absolutely heading into a time of great change and uncertainty. That's inherently unsettling for a lot of people, and the stability that a lot of us felt when we were growing up—I think you're correct that it's not going to be there. We're creating a very, very different world. How do you make a positive impact in that world?

I would say, just think about this broad trend of dematerialization, of doing more with less, and of treading more lightly on the planet, and there are all kinds of ways you can participate in that trend. Another one of the wild things I learned writing the book is that aluminum cans used to be about four or five times heavier than they are now. Just your beer can or your soda can weighed a lot more. Beverage companies and packaging companies worked hard to reduce that weight because the consumer doesn't value it. All I want is my beer, and that aluminum costs money. You can think about it as corporate greed, which it is. At the same time, the cumulative savings are hundreds of thousands of tons of aluminum.

So, by participating in that process, go be a packaging engineer. Go work on CAD/CAM software. Go work on this trend of digitizing our world because then you're also working on dematerializing our world and on treading more lightly on this beautiful planet that we all live on. And if you're an aware and informed citizen, you can help steer governments and companies in the right direction here.

I think there're tons of ways to be a positively involved citizen and consumer here. And, yeah, you should probably also think about having experiences instead of things. The research is pretty overwhelming that more things actually don't make us very happy or satisfied. Go have beautiful experiences.

Jesse Ausubel, whom I mentioned before and whose work I respect greatly, has this great phrase. He says we need to make nature worthless. What he means by that is we need to make it economically worthless. So what if that tree's over there? I don't have any desire to chop it down. I can't make a buck off it. Let me go sit under it and talk to somebody or read a book. He's not saying nature is worthless. He's saying let's make its economic value low as quickly as possible. Amen to that.

Research has revealed a trend of dematerialization in the United States: As the economy is growing, the country is actually using less timber, metals, water, and other resources. And this trend is spreading to other parts of the globe.

- ✓ New technologies and digitization are helping dematerialization happen. As individuals and companies do more online, for instance, they use fewer resources like paper.
- ✓ Developing nations that are building their economy and looking to urbanize must build infrastructure, which traditionally has required vast amounts of resources. But these countries may reach urbanization quicker than expected, since they will follow different technology and materials paths than more prosperous nations.
- ✓ While dematerialization is a promising development, it is not enough to stop climate change. Decarbonization is also needed, and it isn't happening fast enough. Further action, like effective public policy, is required.

Adapted from "Dematerialization and What It Means for the Economy—and Climate Change" on HBR IdeaCast *(podcast), September 17, 2019.*

Section 2

TAKING ACTION

WHY "DEGROWTH" SHOULDN'T SCARE BUSINESSES

by Thomas Roulet and Joel Bothello

The concept of degrowth dates back to the 1970s, when a group of French intellectuals led by the philosopher Andre Gorz proposed a simple idea: In response to mounting environmental and social problems, they suggested that the only real solution was to produce and consume less—to shrink our economies to cope with the carrying capacity of our planet. The proposal was considered by many at the time to be too radical. But with today's climate crisis, debates around degrowth have been reinvigorated, and many major figures such as Noam Chomsky, Yanis Varoufakis, and Anthony Giddens have, to varying degrees, expressed support for the idea.

For others, though—especially business leaders—degrowth is completely unthinkable, not least because of the anticapitalist and anticonsumerist roots of the term. The prevailing view is that growth is an economic necessity, and any threat to that not only undermines business but basic societal functioning. For instance, the chairman and former CEO of H&M Karl-Johan Persson recently warned about the dire social consequences of what he perceives to be a movement of "consumer shaming."[1] Framed in these terms, the resistance of multinational CEOs and entrepreneurs alike is predictable, as is the reluctance of politicians to promote degrowth policies that would potentially prove unpopular with key constituents. The economist Tim Jackson provides a concise assessment: "Questioning growth is deemed to be the act of lunatics, idealists, and revolutionaries."[2]

Critics of degrowth have also put forth other arguments that, at face value, seem valid: the economist Joseph Stiglitz argues, for instance, that since growth is unquestionably good for human development, we simply need a different *kind* of growth that is better for the environment, not less of it.[3] Others argue that the philosophy of degrowth does not seriously account for technological innovation—specifically the idea that we can continue current growth patterns if we innovate products that are less resource intensive and generate fewer waste by-products.

There are, however, problems with these perspectives. First, given the finite nature of our planet, infinite economic growth—even of a different variety—is a logical impossibility. Secondly, innovation and improvements produce, in many cases, unintended consequences. One of these is the Jevons paradox, where individuals compensate for efficiency through increased consumption. For instance, more energy-efficient refrigerators lead to more refrigerators in a home.

The third and most fundamental issue is that the degrowth movement has already begun: at a grassroots level, consumer demand is actively being transformed, despite political and corporate reticence. A recent YouGov poll in France highlights that 27% of respondents are seeking to consume less—double the percentage from two years prior.[4] The number of people eating less meat or giving it up altogether has been rising exponentially in recent years, too. Similarly, the movement of *Flygskam* (literally "flight shaming" in Swedish) has had early successes in reducing pollution: 10 Swedish airports have reported considerable declines in passenger traffic recently, which they attribute directly to Flygskam. In the apparel industry, fast fashion is still popular, but garment manufacturers like H&M are preparing for a backlash as consumers voice growing criticism of the ecological impact of clothing. Accounts such as these indicate how consumers in many contexts are increasingly conscious of the negative consequences of consumerism and are seeking to change their habits. We are witnessing the emergence of consumer-driven degrowth.

These stories also indicate how degrowth opens new opportunities: Some companies and industries will certainly be disrupted, but others that are sufficiently prepared for such transitions will handily outmaneuver their competitors. For instance, Flygskam has been a boon for train travel, bolstered by a social media movement called *Tågskryt* ("train brag"). Meanwhile, reduced meat consumption has been accompanied by an explosion in meat substitutes that produce 1/10th of the greenhouse gases compared to the real thing. Accordingly, degrowth reshuffles competitive dynamics within and across industries and, despite what many corporate leaders assume, offers new bases for competitive advantage.

Based on our examination of companies at the forefront of the degrowth movement, we've identified three of their strategies that can apply to larger incumbent firms.

First, firms can pursue *degrowth-adapted product design*, involving the creation of products that have longer life spans, are modular, or are locally produced. Fairphone, a social enterprise, eschews the built-in obsolescence of larger mobile device manufacturers and produces repairable phones that dramatically extend their longevity. Similarly, the menswear brand Tom Cridland sells high-quality, durable products that run counter to fast fashion principles, including the 30 Year Sweatshirt. Although incumbents have yet to follow suit, such transformations are not without precedent: For example, the American auto industry was forced to move away from planned obsolescence, which was a common practice dating back to the 1920s, when Japanese competitors seized the market in the 1970s and 1980s with more reliable and fuel-efficient vehicles that were built to last.

Second, firms can engage in *value-chain repositioning*, where they exit from certain stages of the value chain and delegate some tasks to stakeholders. As an example, the vehicle manufacturer Local Motors created a proof-of-concept recyclable vehicle crafted with 50 individual parts printed onsite, compared with the roughly 25,000 parts required for a traditional vehicle. The company crowd-sourced designs and crowdfunded the project from their potential consumers. Larger firms such as Lego have also taken advantage of this model, launching marketplaces for either creating new designs or trading used products. This way, the firm creates different ways to consume despite production limits. Firms that incorporate stakeholder engagement in their operations are thereby faster to adapt to degrowth when it becomes more mainstream.

Third, firms can lead through *degrowth-oriented standard setting.* This entails creation of a standard for the rest of the industry to follow. The apparel company Patagonia—which explicitly follows an "antigrowth" strategy—is the poster child for this philosophy, offering a worn-wear store and providing free repairs for not only their own products but also for those of other garment manufacturers. Walmart and Nike have solicited advice from Patagonia on such practices, and more recently H&M imitated the service with a pilot in-store repair facility. In a similar vein, the automobile company Tesla released all its patents in 2014, seeking to catalyze the diffusion of electric vehicles. Such initiatives were not merely marketing ploys but also strategies to standardize a practice or technological platform throughout an industry—one in which companies like Patagonia or Tesla would have existing expertise.

These strategies illustrate potential ways that firms can adapt to consumer-driven degrowth. Firms may pursue more than one strategy (or all three) simultaneously: In 2016, for example, Google attempted to create a longer lasting phone with modular components, soliciting feedback from supply chain actors on how to create standardized parts for their handset. Although "Project Ara" was ultimately canceled, it did reveal a common thread among the strategies. Effective and inclusive communication with stakeholders across the supply chain is crucial, but framing the project in a way that all those individuals can buy into requires considerable effort and adjustment through trial and error.

As we continue to grapple with climate change, we can expect consumers, rather than politicians, to increasingly drive degrowth by changing their consumption patterns. Firms should think in an innovative way about this consumer-driven degrowth as an opportunity, instead of resisting or dismissing the demands of this small but growing movement. Businesses that successfully do so will emerge more resilient and adaptable—instead of necessarily selling more, they will sell *better*, and grow in a way that satisfies consumers while respecting the environment.

Individuals are growing increasingly conscious of the negative consequences of consumerism, leading them to buy less. This consumer-driven degrowth is becoming a concern for companies who see growth as an economic necessity. But three strategies can help:

- ✓ **Degrowth-adapted product design.** Companies can create products that have longer lifespans, are modular, or are locally produced. For example, Fairphone produces repairable phones to extend their longevity.
- ✓ **Value-chain repositioning.** Businesses can exit from certain stages of the value chain and delegate tasks to other stakeholders, including the customers themselves. The vehicle manufacturer Local Motors, for instance, uses crowdsourcing design and crowdfunding to build new products.
- ✓ **Degrowth-oriented standard setting.** Companies can create a standard for the rest of the industry to follow. Patagonia, the poster child for this philosophy, has opened a worn-wear store and provides free repairs to offset those concerned about growth in the apparel industry.

NOTES

1. Hanna Hoikkala, "H&M CEO Sees 'Terrrible' Fallout as Consumer Shaming Spreads," Bloomberg, October 27, 2019, https://www.bloomberg.com/news/articles/2019-10-27/h-m-ceo-sees-terrible-fallout-as-consumer-shaming-spreads.

2. Christiane Kliemann, "Can Companies Do Better by Doing Less?" *The Guardian*, August 1, 2014, https://www.theguardian.com/sustainable-business/2014/aug/01/companies-degrowth-sustainable-business-doing-less.

3. Joseph E. Stiglitz, "Is Growth Passé?" Project Syndicate, December 9, 2019, https://www.project-syndicate.org/commentary/climate-change-demands-transition-to-green-growth-by-joseph-e-stiglitz-2019-12?barrier=accesspaylog.

4. Giulietta Gamberini, "L'envie de 'consommer moins' croît nettement en France," *La Tribune*, September 20, 2019, https://www.latribune.fr/economie/france/l-envie-de-consommer-moins-croit-nettement-en-france-828457.html.

Adapted from content posted on hbr.org, February 14, 2020 (product #H05FC7).

6

IS YOUR COMPANY READY FOR A ZERO-CARBON FUTURE?

by Nigel Topping

There is growing public demand for a rapid transition to a zero-carbon economy. But global protests and youth climate strikes are not enough to create change alone. Companies need to take action. Beyond the very serious threats the current crisis poses to our planet, organizations are increasingly seeing the material risks it poses to their business.

U.S. financial regulator Rostin Behnam likened the financial risks from climate change to those caused by the mortgage meltdown that led to the financial crisis of 2008.[1] And recently, AT&T—which has already lost $847 million to climate disasters—announced that they will be paying the U.S. Department of Energy to track climate-related events that could damage their infrastructure in coming years.

Businesses that bake carbon reduction into their strategies will not only reduce these kinds of risks from affecting their organizations, they will see significant benefits as well: increased innovation, competitiveness, risk management, and growth.

More than 900 global companies representing over $17.6 trillion in market cap are already ensuring that their business strategies are built for growth and emissions reductions through the We Mean Business Take Action campaign. (We Mean Business is a nonprofit coalition of which I am CEO.) This includes over 560 companies that have committed to set ambitious science-based emission reduction targets, and over 175 that have committed to switching to 100% renewable electricity. Beyond that, companies are beginning to use their influence to speed an economy-wide transition by supporting climate policies targeting net-zero emissions by 2050. Others are demanding climate action throughout their supply chains.

Your organization also has a responsibility to become a part of the solution. Failing to do so will impact your ability to attract talent, manage risk, and innovate for growth. Below are a few critical steps you can take to set your business up for success in a zero-carbon future.

Align Your Company with the Paris Agreement

The science has never been clearer. The 2018 IPCC special report on the impacts of global warming of 1.5°C highlights the importance of aligning emission reductions with the goals of the Paris Agreement, and striving for net-zero emissions by 2050—at the latest.[2]

Science-based greenhouse gas emission reduction targets are the gold standard for companies setting emissions reduction goals, both in their direct operations and across their value chains. It is now possible for organizations to set targets that are in line with the level of decarbonization required to limit global warming to 1.5°C. These targets, as ambitious as they are, are vital to reaching net-zero emissions by 2050, and they should be the ultimate goal for all companies.

If you're hesitant, consider the risks of not acting: The world's largest sovereign wealth fund—Norway's $1 trillion Government Pension Fund—confirmed it will divest some $13 billion of fossil fuel–related investments. This is one of many signals that there will continue to be a global move away from fossil fuels, indicating the need for companies to incorporate the emissions impact of their assets into their investment plans or be left with assets that will rapidly lose value. Science-based targets will provide you with a way to future-proof your business plans by ensuring that all strategic decisions incorporate climate risk and opportunity analysis. This will simultaneously drive zero-carbon innovation and help you guard against stranded assets.

To date, the majority of the 560+ companies that have jumped on board report improvements to brand reputation and investor confidence. Consumers and investors are increasingly aware of the effects their choices have on the environment. Companies who commit to these targets, then, are gaining a competitive advantage in multiple areas of their business.

Join a Transformative Initiative

Committing to achieving net-zero emissions by 2050 is no doubt an ambitious goal. There are several initiatives companies can turn to for support.

The Climate Group's global EP100 initiative is a good place to start. It brings together a growing group of energy-smart companies committed to using energy more productively with the goal of lowering greenhouse gas emissions and accelerating a clean economy. As part of the EP100 initiative, companies can commit to doubling energy productivity and to net-zero carbon buildings through the Net Zero Carbon Buildings Commitment.

Companies engaged in this initiative report cost savings as well as emissions reductions. For example, energy productivity improvements at Wisconsin-based Johnson Controls contributed to a 41% reduction in the company's greenhouse gas emissions intensity and over $100 million in annual energy savings.

In addition, collaborative initiatives, like the Low Carbon Technology Partnerships initiative (LCTPi), led by the World Business Council for Sustainable Development, bring companies together to generate shared natural climate solutions across the value chains in specific business sectors. LCTPi focuses primarily on the agricultural, energy, and transport industries. These kinds of initiatives provide businesses with greater access to resources and innovations that can help them develop new markets.

Commit to 100%

Committing to doing something completely—to doing it 100%—leaves no room for excuses and will send a powerful signal to your stakeholders. If you commit to switch 100% of your electricity consumption to renewable sources, as opposed to 20% or even 50%, your objective will be clear to everyone inside and outside of your organization.

Over 175 of the world's most influential companies have already made this commitment through the global corporate leadership initiative, RE100. When they have made the full switch to 100% renewable electricity, these RE100 companies will be generating demand for over 184 terawatt-hours (TWh) of renewable electricity annually, more than enough to power Argentina and Portugal. This is driving up demand for renewable electricity and creating a shift in demand patterns away from fossil fuels across the global power system.

Google, Autodesk, Elopak, and Interface are just a few of the companies that have already achieved their goal and are now powered by 100% renewable energy. Not only are these organizations creating change, they are saving money as the price of wind and solar continues to drop, and they are demonstrating to their stakeholders—including investors, customers, and policy makers—that they see a future in which businesses are powered by renewables.

The same efforts are being made in the transportation sector. With air-quality legislation expected to increasingly restrict polluting vehicles in cities around the world, companies are realizing that it pays to get ahead and make the transition to electric vehicles (EVs). More and more businesses are committing to transition their auto fleets through the global initiative, EV100. LeasePlan, a car leasing company with 1.8 million vehicles on the road, is aiming to transition its employee fleet to 100% EVs by 2021—one step toward their larger goal of reaching net-zero emissions by 2030. In addition to the environmental benefits, their EV integration could dramatically reduce the costs of fleets, as charging an electric vehicle is cheaper than buying gas, and maintenance costs are also lower. Deutsche Post DHL is already seeing 60–70% savings on fuel costs and 60–80% savings on maintenance from its StreetScooter EVs.

Review Your Industry Groups

Industry groups look out for companies' strategic interests and are based on common lines of business. If the groups that your company is a member of are not taking serious action to address the climate crisis, whole industries are at risk of getting left behind once we do reach net-zero emissions. Don't let outdated lobbying positions hold your company back.

The time has come for businesses to review their membership of trade groups and make sure that their climate action goals are aligned. If they aren't, use your company's influence to help change the group's position, or leave the group to show policy makers where you stand. You won't be alone in doing so.

Volkswagen put VDA, the German car maker lobby group, on notice that it will leave unless it adjusts its position on the auto-sector transition and starts supporting EVs. In addition, Shell is walking away from the American Fuel and Petrochemical Manufacturers association over its lack of support for the Paris Agreement. Finally, Unilever CEO Alan Jope has requested that all trade bodies the company is associated with confirm that their lobbying positions on climate are consistent with Unilever's own goals.

Get Smart on Climate Governance

Your plans to tackle climate change will only work if your company has the right governance in place to support them. This includes equipping board and management teams with knowledge and skills that will help them recognize the risks and opportunities posed by the climate crisis.

If you are running a global food company, for example, ask: Is my organization up to date with the findings of the EAT Lancet report? Do we have board expertise on the societal shift away from meat, and is our corporate venturing aligned with this shift? Can we explain to staff and customers how our business model is evolving to protect nature rather than harm it?

To help in this effort, Ceres and The B Team have published a primer on climate competent boards, which also focuses on the adaptability and relevance of the Task Force on Climate-related Financial Disclosure (TCFD) recommendations. These guidelines emphasize the financial risks associated with climate as well as how this informs all business strategy.

Speak Up in Support of Climate Policy

Your company can inspire legislators to create more drastic and ambitious climate policies through face-to-face dialogue. This was highly effective during the negotiations of the Paris Agreement in 2015, when representatives from leading organizations were able to sit down with policy makers and talk openly about the challenges and opportunities different policies would bring to their businesses. The conversation needs to continue.

Many businesses are well positioned to help inform ongoing policy discussions based on their experience with emissions reduction plans. Those that have acted to help improve the state of the climate emergency have the unique ability to point to the progress made through their efforts, demonstrating that climate action is feasible and that inaction is costly.

In Japan, 93 businesses—representing sales of approximately $670 billion and electricity consumption of 36 TWh—called upon the Japanese government to include a goal of net-zero emissions domestically by 2050. Since then, Japan's cabinet has outlined its emissions reduction strategy, which aims to transition the economy to being "carbon neutral" close to that time frame. Hundreds of businesses also called upon the EU to commit to net-zero greenhouse gas emissions by 2050, at the latest. The U.K. government has already announced it is legislating for net-zero emissions by 2050, and pressure is mounting on the EU to follow suit.

Communicate Your Purpose

The more businesses that share the efforts they are making through their reporting and external communications, the more visible they will be to policy makers, customers, and employees. Setting this example can help give those people the confidence they need to increase their own climate ambition and help drive the market shift required to spark competition and innovation.

Perhaps the largest benefit of this is that it has the potential to help put into action long-term climate policies that provide businesses with the clarity they need to decarbonize products and services in faster and smarter ways.

Making your efforts visible to the public will also help your company attract and retain new generations of talent. Some 75% of millennials expect employers to address the climate crisis, and recent research suggests that Generation Z takes an equally strong stance on climate issues.[3]

Companies looking to harness the benefits of climate action need to step up and commit to taking these crucial steps—and don't forget to shout about it when you do. Inspiring others to work toward a zero-carbon future is the best way to drive innovation and ensure that you succeed while others fall by the wayside. We all have a responsibility to tackle the climate crisis and to help drive toward a solution that works for our economies and our planet.

There is an increasing public demand for a zero-carbon economy, and if businesses don't take steps toward folding carbon reduction into their strategies, they could face financial losses and other risks. Follow these steps:

- ✓ **Align your company with the Paris Agreement.** Strive for net-zero emissions by 2050. Use science-based targets and ensure that all strategic decisions incorporate climate risk and opportunity analysis.
- ✓ **Join a transformative initiative.** Find support through projects like the Climate Group's global EP100 initiative, the Net Zero Carbon Buildings Commitment, or the World Business Council for Sustainable Development's Low Carbon Technology Partnerships.
- ✓ **Make a goal of 100%.** Send a powerful signal to your stakeholders by committing to 100% in your environmental initiatives—for instance, by switching *all* of your electricity consumption

to renewable sources, as opposed to only 20% or 50%.

- ✓ **Review your industry groups.** Assess your membership in trade groups and make sure their climate action goals are aligned. If they're not, help change the group's position or leave the group entirely.
- ✓ **Get smart on climate governance.** Equip boards and management teams with knowledge and skills that will help them recognize the risks and opportunities posed by the climate crisis. Read recent environmental reports and regulations, and ask if your company is up to date.
- ✓ **Speak up in support of climate policy.** Inspire legislators to create more drastic and ambitious climate policies through face-to-face dialogue.
- ✓ **Communicate your purpose.** Make your reporting and external communications visible to policy makers, customers, and employees.

NOTES

1. Coral Davenport, "Climate Change Poses Major Risks to Financial Markets, Regulator Warns," *New York Times*, June 11, 2019, https://www.nytimes.com/2019/06/11/climate/climate-financial-market-risk.html.

2. Intergovernmental Panel on Climate Change, "Special Report: Global Warming of 1.5°C," 2019, https://www.ipcc.ch/sr15/.

3. Glassdoor Team, "New Survey Reveals 75% of Millennials Expect Employers to Take a Stand on Social Issues," Glassdoor, September 25, 2017, https://www.glassdoor.com/blog/corporate-social-responsibility/; Kim Parker, Nikki Graf, and Ruth Igielnik, "Generation Z Looks a Lot Like Millennials on Key Social and Political Issues," Pew Research Center, January 17, 2019, https://www.pewsocialtrends.org/2019/01/17/generation-z-looks-a-lot-like-millennials-on-key-social-and-political-issues/.

Adapted from content posted on hbr.org, June 21, 2019 (product #H050QH).

MAKING CRYPTOCURRENCY MORE ENVIRONMENTALLY SUSTAINABLE

by Marc Blinder

Blockchain has the power to change our world for the better in so many ways. It can provide unbanked people with digital wallets, prevent fraud, and replace outdated systems with more efficient ones. But we still need this new and improved world to be one that we want to live in. The largest cryptocurrencies—Bitcoin, Bitcoin Cash, and Ethereum—require vast amounts of energy consumption to function. In 2017, blockchain used more power than 159 individual nations, including Uruguay, Nigeria, and Ireland.[1] Unsurprisingly, this is creating a huge environmental problem that poses a threat to the Paris climate-change accord.

It's a brutal, if unintended, consequence for such a promising technology, and "mining" is at the heart of the problem. When Bitcoin was first conceived nearly a decade ago, it was a niche fascination for a few hundred hobbyists, or "miners." Because Bitcoin has no bank to regulate it, miners used their computers to verify transactions by solving cryptographic problems, similar to complex math problems. Then, they combined the verified transactions into "blocks" and added them to the blockchain (a public record of all the transactions) to document them—all this, in return for a small sum of Bitcoin. But where a single Bitcoin once sold for less than a penny on the open market, it now sells for nearly $7,000, and around 200,000 Bitcoin transactions occur every day. With these numbers increasing, so has the incentive to create cryptocurrency "mines"—server farms now spread across the world, often massive. Imagine the amount of energy consumed by 25,000 machines calculating math problems 24 hours a day.

Beyond the environmental concerns, this inefficiency threatens blockchain as a meaningful platform for enterprise. The high energy costs are baked into the system, and, because the cost of running the network is passed on in transaction fees, users of these networks end up paying for them. Initially, companies that use Bitcoin may not see the financial consequences, but as they scale, the costs could become fatal.

The good news: There are various alternatives available that can help organizations cut massive energy costs. Right now, they aren't being adopted quickly enough. Companies that want to keep their heads above water—along with everyone else's—need to educate themselves. Below are two areas that are a good place to start.

Green Energy Blockchain Mining

An immediate fix is mining with solar power and other green energy sources. Each day, Texas alone receives more solar power than we need to replace every non-solar power plant in the world. There are numerous commercial services for powering cryptomining on server farms that only use clean, renewable energy. Genesis Mining, for instance, enables mining for Bitcoin and Ethereum in the cloud. The Iceland-based company uses 100% renewable energy and is now among the largest miners in the world.

We need to incentivize green energy for future blockchains, too. Every company that uses blockchain also defines its own system for miner compensation. New blockchains could easily offer miners better incentives, like more cryptocurrency, for using green energy—eventually forcing out polluting miners. They could also require all miners to prove that they use green energy and deny payment to those who don't.

Energy-Efficient Blockchain Systems

While Bitcoin, Bitcoin Cash, and Ethereum all depend on energy-inefficient cryptographic problem solving known as "proof-of-work" systems to operate, many newer blockchains use "proof-of-stake" (PoS) systems that rely on market incentives. Server owners on PoS systems are called "validators"—not "miners." They put down a deposit, or "stake" a large amount of cryptocurrency, in exchange for the right to add blocks to the blockchain. In proof-of-work systems, miners compete with each other to see who can problem solve the fastest in exchange for a reward, taking up a large amount of energy. But in PoS systems, validators are chosen by an algorithm that takes their "stake" into account. Removing the element of competition saves energy and allows each machine in a PoS system to work on one problem at a time, as opposed to a proof-of-work system, in which a plethora of machines are rushing to solve the same problem. Additionally, if a validator fails to behave honestly, they may be removed from the network—which helps keep PoS systems accurate.

Particularly promising is the delegated proof-of-stake (DPoS) system, which operates somewhat like a representative democracy. In DPoS systems, everyone who has cryptocurrency tokens can vote on which servers become block producers and manage the blockchain as a whole. However, there is a downside. DPoS is somewhat less censorship resistant than proof-of-work systems. Because it has only 21 block producers, in theory, the network could be brought to a stop by simultaneous subpoenas or cease and desist orders, making it more vulnerable to the thousands upon thousands of nodes on Ethereum. But DPoS has proven to be vastly faster at processing transactions while using less energy, and that's a trade-off we in the industry should be willing to make.

Among the largest cryptocurrencies, Ethereum is already working on a transition to proof-of-stake, and we should take more collective action to hasten this movement. Developers need to think long and hard before creating new proof-of-work blockchains because the more successful they become, the worse ecological impact they may have. Imagine if car companies had been wise enough several decades ago to come together and set emission standards for themselves. It would have helped cultivate a healthier planet—and preempted billions of dollars in costs when those standards were finally imposed on them. The blockchain industry is now at a similar inflection point. The question is whether we'll be wiser than the world-changing industries that came before us.

The largest cryptocurrencies require vast amounts of energy consumption to function. In 2017, blockchain used more power than 159 individual nations, including Uruguay, Nigeria, and Ireland. This is creating a huge environmental problem that poses a threat to the Paris Agreement.

- ✓ With the value of cryptocurrency increasing, so has the incentive to create "mines"—massive server farms spread across the world, calculating math problems 24 hours a day.

- ✓ A variety of alternatives are available that can help organizations cut massive energy costs associated with blockchain, but they aren't being adopted quickly enough.
- ✓ An immediate fix is mining with solar power and other green energy sources. Every company that uses blockchain also defines its own system for miner compensation. New blockchains could easily offer miners better incentives for using green energy.
- ✓ Blockchains that use a proof-of-stake system rather than a proof-of-work system are less energy intensive. If the blockchain industry can make the hard choice to adopt greener decisions now, it may avoid having green standards forced upon it in the future.

NOTE

1. "Bitcoin Energy Consumption Index," Digiconomist, https://digiconomist.net/bitcoin-energy-consumption.

Adapted from content posted on hbr.org, November 27, 2018 (product #H04O38).

CLIMATE CHANGE IS GOING TO TRANSFORM WHERE AND HOW WE BUILD

by John D. Macomber

As fires, floods, and droughts increasingly threaten homes, businesses, and other institutions, climate risk has become a financial risk. A National Bureau of Economic Research paper recently concluded that mortgages written on homes in exposed locations are being shed by banks and absorbed by Fannie Mae and

Freddie Mac, government-backed mortgage guarantors.[1] This implies that homeowners and investors have been making location decisions without properly pricing the cost of potential peril, and that the government has been enabling the oversight. Some are even warning that this market failure could lead to a repeat of the 2008 financial crisis, which was also triggered by bad mortgages.

It's not just homeowners investing recklessly—many businesses have been equally shortsighted in where they place new assets, such as factories, and what to do with existing assets in once-safe areas now threatened by these perils. While laudatory efforts continue to mitigate climate change at the international level, it's long past time to accept that the climate is already irreversibly changing, and we must adjust our mindset accordingly. We can't just keep piling sandbags, pumping basements, dousing flames, and expecting government bailouts forever; a methodology is needed for homeowners, businesses, mortgage holders, governments—all of society—to figure out which assets to reinforce and what other courses of action are available.

Alas, we seem to be headed in the wrong direction. While virtually no private insurance companies retain residential flood risk in Florida, Virginia, and other coastal states due to sea rise, government-subsidized programs such as the National Flood Insurance Program (NFIP) continue to keep residences insured. Making things even worse is a general lack of restrictions on where one can build and what can be built. A society that prides itself on free will and self-determination is loath to say what a property owner can and cannot build on their own land as long as it meets rudimentary building and zoning codes, so more and more people move into harm's way in flood plains, low-lying coastal areas, and tinder-dry Western landscapes.

Why this dysfunction? To begin with, all three of these factors—mortgages, insurance, and land development policies—are by design backward looking. They rely on histories of lending defaults, insurance claims, and floods and fires to make determinations. This is logical; there is plenty of hard empirical data to draw from, and a switch to looking forward would risk becoming a matter of speculation over the merits of one modeling method or another. There are also plenty of market incentives to keep housing prices high. But the backward-looking approach seems overwhelmed today as climate events—and follow-on effects on the ground—are moving so fast.

There is a better way. My research and that of others indicates that there are five basic choices in investing in resilience: reinforce, retreat, rebound, rebuild, and restrict.[2] Together, they can be used as a decision-support tool for what to do with assets exposed to climate risk.

Reinforce

This is often the default recommendation in the face of climate-related perils. But only in some circumstances is it is the *right* reaction. A good example is Texas Medical Center (TMC) in Houston. Following severe damage from Hurricane Allison in 2001, TMC invested hundreds of millions of dollars in resilience improvements, including flood bulkheads, sky bridges, elevation of electrical equipment, and more. When Hurricane Harvey hit in 2017, Houston was soaked by unthinkable amounts of rain and the regional flooding persisted for weeks—but with its resilience interventions in place, Texas Medical Center hardly missed a beat. Investing in reinforcement made sense in this instance because, first, the direct and indirect costs of being hit were huge; the hospital had the ability to raise the capital even though the investment in resilience was not sure to pay off (since there might have never been another big storm); and, second, because the incremental capital spending relative to the entire balance sheet was manageable. But not every potential investment in resilience makes economic sense. We can't pay for a blanket policy to resist sea rise and rainfall and fire in every situation forever, regardless of cost/benefit. When should a different route be taken?

Retreat

Consider the opposite scenario. After several consecutive years of being overwhelmed by riverine flooding, multiple small shopkeepers and restaurants in Ellicott City, Maryland, collectively abandoned the historic downtown for higher, dryer land. This was the right course of action for them: The cumulative flooding had a substantial cost relative to their balance sheets; the price to reinforce (by raising floors or building sea walls) was too high, and resources were thin.

Rebound

In Miami and Miami Beach, most new commercial or condo buildings on the water are built with extra-high first floor elevations ("freeboard"), with expensive equipment located out of harm's way on the second floor or higher, and with furnishings and materials at ground level and in the basement that can survive temporary seawater inundation, be pumped out, be cleaned quickly, and then put back in service. The famous wooden plank walkways on St. Mark's Square in Venice are similar: They embrace neither reinforcement or retreat, but rather are focused on "living with water"—how to rebound in a cost-effective way.[3]

Rebuild

This is essentially the path chosen by the homeowners in Houston, Texas, or the Hampton Roads area in Virginia, where FEMA has paid to rebuild thousands of homes multiple times each. Rebuilding is a great choice—particularly if you are rich or if you can use someone else's money. The long-term questions, though, are: Will there always be that money available? Will it go to those who are most in need or to help higher-income individuals, as reported by *U.S. News and World Report*?[4] And, of course, how long will the other taxpayers support aiding those in danger zones? In recent years, Congress has authorized about $200 billion per year for disaster relief in California, Iowa, Texas, Florida, New York, Puerto Rico, and elsewhere (on top of subsidizing the NFIP). It's not clear that, in a time of deficits, this largesse will be perpetual.

Restrict

This is a strategy that the private sector mortgage and insurance companies appear to already be following—

they are not investing in harm's way and, based on press reports, they are drawing back. Some other large asset owners such as real estate investment trusts or retailers thinking about store locations are avoiding purchasing or building properties in high-risk areas. But there is a public policy angle, as well: Recently the utility PG&E was forced to pick up the cost of property damage from California wildfires. A strong argument could be made that the local government should have restricted these structures from being built in the first place.

In the coming years, growth pressures will increase in many attractive locations (such as coastal cities and Western states). Americans will be forced to make some tough decisions. Subsidized mortgages and artificially cheap insurance have let us put off the hard reckonings a little longer, but if a severe price correction (and subsequent economic shock) is to be avoided, all asset owners in potentially exposed locations will need to pick one of these five paths for investing in resilience. The right choice in each situation depends on circumstances, the level of exposure, the cost of potential damages, the cost to reinforce, and resources available. The challenge for homeowners, investors, mayors, and all of us is to look ahead, not behind, and to make these choices with intent—before circumstances take the choices out of our hands.

Businesses need to think about where they are placing new assets and what to do with existing ones that are now threatened by extreme weather events. Five basic choices in investing in climate resilience can be used to help you decide what to do with assets exposed to climate risk:

- ✓ Reinforce building structures through flood bulkheads, sky bridges, or high walls to protect against extreme events.
- ✓ Retreat by moving assets and people away from climate-endangered areas.
- ✓ Rebound by introducing building features designed to deal with damage—for instance, lower levels that can deal with flooding—without hurting the overall structure.
- ✓ Rebuild after disaster, using government aid and other funding.

✓ Restrict building by avoiding purchasing or building properties in high-risk areas or through strict zoning laws.

NOTES

1. Amine Ouazad and Matthew E. Kahn, "Mortgage Finance in the Face of Rising Climate Risk," NBER Working Paper 26322, September 2019.

2. Regional Plan Association, "Where to Reinforce, Where to Retreat?" Fourth Regional Plan Whitepaper, May 22, 2015. https://www.rpa.org/publication/fourth-regional-plan-whitepaper-where-to-reinforce-where-to-retreat.

3. Urban Land Institute, "The Urban Implications of Living with Water," June 19, 2019, https://boston.uli.org/uli-resources/the-urban-implications-of-living-with-water/.

4. Cecelia Smith-Schoenwalder, "Study: FEMA Flood Buyouts Favor the Wealthy," *U.S. News and World Report*, October 9, 2019, https://www.usnews.com/news/national-news/articles/2019-10-09/study-fema-flood-buyouts-favor-the-wealthy.

Adapted from content posted on hbr.org, October 16, 2019 (product #H0583W).

HOW BUSINESSES CAN BRACE FOR CATASTROPHE

by Yvette Mucharraz y Cano

On Sept. 19, 2017, a severe earthquake struck Mexico City, killing 370, injuring thousands, and destroying scores of buildings. For a short time, the city ground to a halt as rescue efforts took priority over normal commerce. But then something curious happened: Some businesses were able to return to work relatively quickly, while others lagged behind or even failed to recover and disappeared.

The earthquake provided a natural laboratory to study a crucial but often elusive business capacity: organizational resilience. Two years after the disaster, I met with a group of leaders whose businesses survived the quake to understand what set them apart. I found that their answers fit nicely with a leading theory about organizational resilience: that it falls into successive stages—anticipation, coping, and adaptation.[1]

Anticipation

Resilience starts with the awareness generation of potential risks to the organization, supported by data collected through risk assessments and experience. Resilient organizations put in place the necessary action protocols, business continuity efforts, financial provisions, human resources, technology, and infrastructure to ensure they can respond well to high-consequence, low-probability events. For instance, a consulting services company that lost a building in the center of the city and did not have human casualties developed in advance the emergency processes and protocols that protected their clients and employees when the earthquake hit because they evacuated easily and knew how to act. Also, the leaders acquired the insurance coverage that would protect the organization from a catastrophic incident, and the organization received the benefits almost immediately to establish a new office in a few weeks.

Specifically, the businesses I met with undertook several steps to help prepare them in advance for unforeseen catastrophes:

They built a strong organizational culture

The corporate culture serves as a preparedness mechanism for environmental events, especially when it is supported by a strong commitment from organizational leaders with ethics and values and when trust is present in the organizational community. For instance, in another organization in which the building collapsed only 15 seconds after the earthquake started, the team and the group of leaders who were interviewed shared a common view about how they lived their corporate values of empowerment, innovation, teamwork, integrity, and inclusion. In the recovery phase, the leaders relied on the team members to make decisions and create solutions to continue operating while the business partners looked for financial support, addressed the legal issues, and supported the families of the affected employees.

They promoted emergent leadership

Organizations' emergency responses are activated by emergent leaders during crises, and they are tested during minor crises. Emergent leadership is observed in the response to extreme events and is associated with organizations that have low power distance, distributed leadership, flexible structures, and processes that reflect empowerment and delegation of authority.

They invested in safe and secure workspaces

The concept of "leadership through place" is critical in environmental hazards that affect organizations, as the loss of the space—for example, a building—impacts individuals deeply, and the feeling of being lost may be relieved by redefining the organization's identity to foster a sense of belonging and create a "new normalcy" in a new space.

Consequently, the selection of the space in which an organization operates is more than an incidental aspect. Leaders must provide a safe environment to preserve physical integrity. Adherence to official standards and norms is critical to protect individuals who attend an organizational setting, including employees, contractors, and visitors. Facilities must be evaluated by construction and civil protection experts and secured by insurance coverage. Additionally, employees must be covered by health and life insurance policies and social security in compliance with labor and civil protection laws.

Leaders must be aware and accept that the threat of an environmental hazard is continuously latent. Action is therefore required on a daily basis. Evacuation drills are not enough; for instance, as part of the continuity of business strategies, it is critical to identify alternate sites from which to operate and to simulate the communication flow, as there may be a suspension of services such as Wi-Fi or social media during an emergency due to a lack of power and the saturation of communication networks. It is also necessary to verify that security measures such as bars or locks installed on access doors or emergency exits do not become obstacles to evacuation during an emergency. In the case of organizations that have animals in their facilities, they also need to be part of the emergency protocols. Finally, critical prevention decisions include the location of company servers, official documents, or objects difficult to replace (for example, art) and cash vaults to secure them in case of an emergency.

Technology can make a difference

One of the primary considerations that leaders mentioned repeatedly was the use of information and technology to back up organizational data and provide connectivity as a precautionary measure to preserve information. This is a critical intangible asset for the organization and for continuity of business purposes.

Coping

To cope with an environmental crisis effectively, inter-organizational and multisectoral coordination is essential—among governmental institutions, the private sector, universities, nongovernmental entities, and civil society. This coordination can become a source of solidarity and a potential source of relief during a crisis. Resilience must be cultivated cross-functionally, from the prevention stage, activated during the event, and in the recovery phase. For instance, an alliance known as ARISE—the private-sector alliance for disaster-resilient societies led by the UN office for Disaster Risk Reduction (UNDRR)—was established in 2018 in Mexico. Though a global initiative, its Mexican operation fosters the participation of the business community, governmental entities such as the National Coordination of Civil Protection, and other nongovernmental institutions. The ARISE alliance is focused on building resilient societies by disaster prevention and risk reduction.

After the earthquake, there were exemplary organizations that worked actively in the emergency response efforts. As an illustration, an organization in the south of Mexico City dedicated time and energy to the rescue efforts of the neighborhood. One of the leaders became the leader of a donations center and coordinated the support mechanisms with public entities. Another private company, dedicated to rebuilding damaged buildings, has participated in an ongoing initiative with governmental institutions to support the reconstruction process.

Furthermore, the role of social media is fundamental during the immediate response stage, as it is a potential source of aid that activates social capital and is directly related to the reduction of human casualties. An example of this coordination was the emergence of #Verificado19s, a digital platform that verified social media information with civil society participation and managed data to make the citizens' response more efficient after the 2017 earthquake.

Adaptation

Address the grief that results from the crisis

Emotional support by leaders is essential following traumatic events. The socialization of the grieving process in the corporate community is a means of promoting this emotional response and may lead to posttraumatic growth. Resilience in this context is not only a return to normalcy after a disaster but also involves leaders' acknowledgment of the grieving process to mourn for what was lost. Companies are more than their assets; at their core are human beings who collaborate, and they should be in the center.

For instance, a rescuer who was interviewed for the research mentioned that every day during the emergency response following the earthquake, the group of rescuers formed a circle to share their emotions and tell the day's stories in a safe community as a way to prevent posttraumatic stress. Similar rituals should be introduced into corporate settings as well.

Learn and grow from adversity

Posttraumatic growth in an organization is related to the organization's adaptability and is promoted by a learning process in which leaders reflect on what they did well after an environmental incident. Resilient leaders are capable of sharing their vulnerability, are not afraid to look for emotional support, and let themselves be guided by their teams when needed.

Post-disaster resilience is an adaptive capacity to grow and develop that does not occur by chance. As has been evident in the years after the earthquake in Mexico, resilience is developed collaboratively. Resilient organizations demonstrated responsibility and solidarity with their employees and with the community and were supported by strong internal and multisectoral networks. These networks were developed primarily in the anticipation stage, and the return on the investment to build them has been the survival of the organization despite the critical nature of the emergency. The enemies of resilience were corruption, conformity with the status quo, a lack of flexibility, and the inability to accept the crisis before it escalated.

In the face of catastrophe, companies need a crucial but often elusive business capacity: organizational resilience.

By looking at how businesses responded to the devastating earthquake in Mexico City, all leaders can learn how to build such resilience before, during, and after a crisis:

- ✓ **Anticipation:** In advance of potential events, build a strong organizational culture that's supported by leaders, promote emergent leadership during crises, invest in safe and secure workspaces, and use information and technology to back up data and provide connectivity.
- ✓ **Coping:** Encourage coordination among governmental institutions, the private sector, universities, and other sectors. Such alliances can then become a source of solidarity and disaster relief during a crisis.
- ✓ **Adaptation:** Acknowledge the emotions employees are feeling, like grief, and encourage learning and posttraumatic growth. Such resilience is developed collaboratively, with employees, the community, and other networks built before disaster strikes.

NOTES

1. Stephanie Duchek, "Organizational Resilience: A Capability-Based Conceptualization," *Business Research* 13 (January 2019): 215–246.

Adapted from content posted on hbr.org, February 6, 2020 (product #H05EE7).

YOUNG PEOPLE ARE LEADING THE WAY ON CLIMATE CHANGE, AND COMPANIES NEED TO PAY ATTENTION

by Andrew Winston

In one of the many oddities of biology, kids hear differently than the rest of us. There are frequencies that only teens and young adults can make out. Lately it seems that the under-20 crowd is hearing one particular high pitch much better than the rest of us, including most business leaders: the alarm that climate scientists have been sounding.

Consider the young Swede, Greta Thunberg. At age 15, Thunberg stopped going to school to protest inaction on climate change, saying there was little point in studying for a future that may not exist. Within months, Thunberg urged immediate action from business leaders at the World Economic Forum and told the UN's secretary general and others at the global climate summit in Poland that they are "stealing [childrens'] future in front of their very eyes."[1] What she started is growing, and she's been nominated twice for a Nobel Peace Prize for her efforts.

In 2018, thousands of Belgian youth marched weekly on the EU capitol of Brussels. And on March 15, 2019, in what may be the largest youth-led protest in history, an estimated 1.6 million students in 300 cities around the world walked out of school to march for climate action. I went to the New York march, and the energy was electric—and I didn't even take it personally when a group of teens called some colleagues and me "old people who need to do something."

There's more: The youth group the Sunrise Movement recently held a somewhat contentious meeting with Sen. Dianne Feinstein of California about her support for climate policies. And a group of teens has sued the U.S. government for failing to protect them from climate change. Younger politicians are making their voices heard, too. Consider what Rep. Alexandria Ocasio-Cortez of New York achieved in just a few months in office. By pushing a broad set of climate and inequality goals under the banner of a "Green New Deal," the youngest woman ever elected to the U.S. Congress has moved the terms of the climate debate significantly.

Before writing this off as a lot of noise, consider the role of youth in previous social movements. Baby Boomers, when they were kids and teens, led the antiwar movement. The famed Greensboro lunch counter sit-in was led by four young men aged 17, 18, and 19. African-American kids bravely desegregated schools, and the first person to get arrested for refusing to give up her bus seat was *not* actually Rosa Parks, but 15-year-old Claudette Colvin. A generation later, Gen X and then Millennials shifted the debate on LGBT rights and gay marriage at a remarkable pace. In fact, it's hard to think of any substantial social movement that *didn't* have young, fearless people at the center.

And now, with the powerful tools of social media and 24-7 connectivity, the pace of social movements is quickening. The "Parkland Teens," the survivors of the horrific school shooting in Florida, attracted millions of Twitter followers in days. Within just a few weeks, they called for marches, for which over a million people showed up around the world. Cut to a year later, and the U.S. House of Representatives passed the first real gun control legislation in many years.

Will this climate movement end up as significant as the antiwar, civil rights, and gay rights movements? It's hard to predict. But what's clear is that we're in the middle of a major realignment of values around climate. It's now unacceptable to young activists, and the millions of people they inspire, to espouse climate denial or play the "let's go slow" card. They don't appreciate being handed a disaster movie for them to live with for 70 to 80 years.

This brings me to business, and a warning: No organization can avoid values shifts. Remember, there were moments in history where it was generally acceptable to use slave labor or children in supply chains, to wink at rampant sexual harassment in offices, and to freely dump pollution in rivers and the air. None of these problems are eliminated today, but very few in business would suggest that they're OK. Morals changed, and then laws.

And while executives do increasingly seem to be moving toward action on climate change, with public pronouncements to cut their own emissions or buy renewable energy becoming the norm in large companies, it's not clear whether those actions are enough to satisfy this next generation of customers and employees. In fact, companies seem to be more comfortable taking public stands on issues like race, immigration, gun violence, and transgender rights *before* speaking strongly on the environment.

But that needs to change now. It's time, in the words of U.S. senator Sheldon Whitehouse, for "corporate good guys" to "show up in Congress to lobby for climate action."[2] We need CEOs in the halls of power at the state and federal level pushing for aggressive policy.

This isn't a new idea, of course, but the history on climate lobbying is sparse. There are "D.C. visit" days organized by a few focused NGOs, and they're always hoping for bipartisan climate solutions. But in reality, with a few exceptions, only smaller companies have been willing to put themselves out there. The big guys sign on to public statements like "We Are Still In," which is a good start but is inadequate to the level of change required. They need to put some skin in the game and become more vocal and more aggressive.

In practice, this will mean disagreeing with politicians, up to and including the president, who say it's too expensive to act, or that climate change is a hoax. In fact, a recent survey shows that 76% of Americans *want* companies to take a stand for what they believe, even if it's politically controversial.[3]

It may just take the youngest Americans to get companies to take a real and public stand for aggressive global action on climate change; after all, if they don't, they risk getting out of step with an entire generation of employees and customers.

Young people are increasingly sounding the alarm about climate change. While many individuals may write this off as noise, businesses need to realize that if they don't pay attention, they might become out of step with an entire generation of employees and customers.

- ✓ This isn't the first time a youth movement is leading the charge in social issues. Baby Boomers led the antiwar movement when they were kids and teens. African American kids bravely desegregated schools, and Gen X and Millennials shifted the debate on LGBT rights and gay marriage.
- ✓ Young activists and those they inspire find it unacceptable to deny climate change or move slowly, yet companies seem to be more comfortable taking a public stand on key social issues like race,

immigration, gun control, and transgender rights *before* speaking strongly on the environment.

✓ While executives and organizations do seem to be moving toward taking climate actions, like buying renewable energy or cutting down their emissions, these actions alone may not satisfy these younger individuals. Executives must take a public stand and aggressively lobby on climate issues—even if it means disagreeing with powerful politicians.

NOTES

1. John Sutter and Lawrence Davidson, "Teen Tells Climate Negotiators They Aren't Mature Enough," CNN, December 17, 2018, https://www.cnn.com/2018/12/16/world/greta-thunberg-cop24/index.html.

2. Carolyn Fortuna, "'Time to Wake Up'—RI Senator Whitehouse Offers Solutions to Climate Change," CleanTechnica, February 27, 2019, https://cleantechnica.com/2019/02/27/time-to-wake-up-ri-senator-whitehouse-offers-solutions-to-climate-change/.

3. Julie Hootkin and Tanya Meck, "Call to Action in the Age of Trump: Business & Politics: Do They Mix?" Global Strategy Group, 5th Annual Study, 2018, https://www.globalstrategygroup.com/wp-content/uploads/2018/02/BusinessPolitics_2018.pdf.

Adapted from content posted on hbr.org, March 26, 2019 (product #H04UZT).

About the Contributors

ITZHAK (ZAHI) BEN-DAVID is the Neil Klatskin Chair in Finance & Real Estate at the Fisher School of Business, The Ohio State University.

MARC BLINDER is chief product officer of blockchain-based start-up AIKON. After earning a political science degree from Princeton and working in Bay Area politics, he developed a social network called MobilePlay (sold to Good Technology) and held leadership roles at Context Optional and Efficient Frontier (sold to Adobe). Follow him on Twitter @mblinder.

JOEL BOTHELLO is an assistant professor in management at the John Molson School of Business, Concordia University. He works at the intersection of organization theory and resilience. Follow him on Twitter @j_bothello and visit his website at joelbothello.com.

ADAM CONNAKER is senior associate responsible for Innovative Finance at the Rockefeller Foundation in New York City, New York.

DANTE DISPARTE is the vice chairman of the Libra Association and the founder and chairman of Risk Cooperative. He is coauthor of the book *Global Risk Agility and Decision Making* and serves on the National Advisory Council of the Federal Emergency Management Agency, FEMA.

STEFANIE KLEIMEIER is professor of entrepreneurial finance and banking at the Open University in the Netherlands. She also holds the positions of associate professor of finance at the Maastricht University in the Netherlands and professor extraordinary at the University of Stellenbosch Business School in South Africa.

JOHN D. MACOMBER is a senior lecturer in the finance unit at Harvard Business School. He teaches in HBS's Business and Environment and Social Enterprise Initiatives.

SAADIA MADSBJERG is managing director at the Rockefeller Foundation in New York City, New York, and coauthor of the *Foreign Affairs* article "The Innovative Finance Revolution."

ANDREW McAFEE is the codirector of the Initiative on the Digital Economy in the MIT Sloan School of Management. He is the author of *Enterprise 2.0* and the coauthor, with

Erik Brynjolfsson, of *The Second Machine Age*. His latest book is *More from Less: The Surprising Story of How We Learned to Prosper Using Fewer Resources—And What Happens Next*.

YVETTE MUCHARRAZ Y CANO is a professor at Ipade Business School.

CURT NICKISCH is a senior editor at *Harvard Business Review* where he makes podcasts and cohosts the podcast *HBR IdeaCast*. He earned an MBA from Boston University and previously reported for NPR, *Marketplace*, WBUR, and *Fast Company*. He speaks *ausgezeichnet* German and binges history podcasts. Find him on Twitter @ CurtNickisch.

THOMAS ROULET is a senior lecturer in organization theory at the Judge Business School and a fellow of Girton College, both at the University of Cambridge. He has provided sociological analyses on different aspects of Brexit in various media outlets (*the Telegraph, l'Humanité, Die Zeit*). Follow him on Twitter @thomroulet.

NIGEL TOPPING serves as CEO of We Mean Business coalition, which harnesses business leadership to drive the innovations and policies that accelerate action on climate change. Previously, Nigel was executive director of CDP (formerly the Carbon Disclosure Project), and he has 18 years of experience in the manufacturing sector.

MICHAEL VIEHS is associate director at Hermes Investment Management in London. In this position, he is responsible for the integration ESG and engagement information in the investment strategies of all public market equity and credit funds. Michael is also an honorary research associate at the Smith School of Enterprise and the Environment at the University of Oxford.

ANDREW WINSTON is the author, most recently, of *The Big Pivot*. He is also coauthor of the best-seller *Green to Gold* and the author of *Green Recovery*. He advises some of the world's leading companies on how they can navigate and profit from environmental and social challenges. Follow him on Twitter @AndrewWinston.

Index

Note: Page numbers followed by *f* indicate figures.